AF725469

Fired Clay in Four Porcelain Clusters

A Comparative Study of Energy Use, Production/Environmental Ecology, and Kiln Development in Arita, Hong Kong, Jingdezhen, and Yingge

Tai Wei Lim

University Press of America,® Inc.
Lanham • Boulder • New York • Toronto • Plymouth, UK

Copyright © 2014 by University Press of America,® Inc.
4501 Forbes Boulevard, Suite 200, Lanham, Maryland 20706
UPA Aquisitions Department (301) 459-3366

10 Thornbury Road, Plymouth PL6 7PP, United Kingdom

All rights reserved
Printed in the United States of America
British Library Cataloguing in Publication Information Available

Library of Congress Control Number: 2014943550
ISBN: 978-0-7618-6428-8 (cloth : alk. paper)—ISBN: 978-0-7618-6429-5 (electronic)

∞™ The paper used in this publication meets the minimum requirements of American National Standard for Information Sciences Permanence of Paper for Printed Library Materials, ANSI/NISO Z39.48-1992.

Contents

Illustrations

FIGURES

TABLES

Acknowledgments

I wish to acknowledge the following people who have given support, encouraged or assisted in the project (in alphabetical order): Colleagues at the Department of Japanese Studies Chinese University of Hong Kong (CUHK); Curators and Friends from the Kyushu Ceramics Museum; Faculty members of Saga University and Kyushu University; editing and other assistance from Ms Helen Chan and Ms Elim Wong; Friends from Kagoshima; Friends/colleagues in the IARC (International Affairs Research Center); Hop Wai grant; Potters in Hong Kong based in Fo Tan, Wan Chai; Friends in the Town of Arita, Imari and Izumi; Friends in Yingge and Taipei; Student helpers of JAS; The Fourth generation of contemporary potters in HK; The Government of Saga Prefecture staff members; The Libraries and Art Museum of CUHK; The South China Programme grant; The Staff Members of Fukagawa and Koransha. I also acknowledge the kind permissions of the Editors of Journal of Asian History (for the rights to reproduce Chapter 2 in this publication), Touyou Bunka Kenkyuu Journal of Cultural Studies (for the rights to reproduce Chapter 3) and Sino-Japanese Studies (for the rights to reproduce Chapters 1 and 4 of this volume). This volume is dedicated to all contemporary potters, ceramicists, porcelain makers in the world and many others who have encouraged and inspired this project.

Introduction

The genesis of this project started with an interest in adapting micro-studies to examine how energy use evolved with technological advancements and changing social norms such as views on environmentalism. This project is made possible with a combination of several research grants: The South China Programme grant which focused on the study of *mingei* (folk) pottery amongst potter communities in Hong Kong and Japan and the Hop Wai grant which focused on the Arita and Jingdezhen comparative case studies. Three research themes consistently ran through my research work. First, how did the use of energy fuels impact on the developmental history of the porcelain-making industry? How did it impact on design concepts, manpower use, environmental conservation and changes in productivity in the porcelain-making industry? The second research theme was how energy use depends on a network of specialists and technicians in any ceramics-making artistic cluster. The choice of fuels in kilns alone is insufficient to support activities that generate revenues for the industry. Energy use and kiln facilities also cannot fully explain the brand recognition factors of Arita and Jingdezhen-made porcelains in the world. This is related to how ceramicist communities in the world organize themselves institutionally to maximize productivity and effectiveness in porcelain distribution. Third, at a cognitive level, I was interested to study changes in production and design thinking about production, environmental integrity, energy use and aesthetic trends amongst the ceramicists and consumers of the porcelain products.

Before embarking on my fieldwork which took me to four territories and states (Japan, China, Hong Kong Special Administrative Region (HKSAR) and Taiwan) and cities and towns (Arita, Izumi, Hong Kong and Yingge art clusters) in these four locations, I carried out interpretive work, textual analyses and library research. I have discussed relevant existing literatures in each of the subsequent chapters, including a few important volumes related to this subject matter. One important publication that I would like to specifically highlight is Brian Moeran's publication *Folk Art Potters of Japan* in this introduction.[1] It makes at least three important contributions to the discussion on Japanese ceramics industries. First, it is one of the few academic literatures found in the market which studies folk pottery (*mingei*) production in a detailed manner within a local community. In my study, I did not restrict myself to *mingei* because my textual research (confirmed by informants later) reveals that *mingei* tends to have limited circulation and distribution and it appeals to a certain socioeconomic class. However, it is an important genre of ceramics that I have also included in my study as a comparative category with other kinds of porcelain encountered in the fieldwork. My study is restricted to high-fired ceramics (mostly fired at 1200 degrees Celsius) as lower-fired ceramics pieces have some qualities that differentiates them from hi-fired porcelain, e.g., greater porosity, less hard and glass-like properties, and typically not associated aesthetically with bone china texture. More importantly, high-fired porcelain wares require more advanced kiln technologies and energy use which are the focus of attention in this volume. In addition, high-fired porcelain wares appeals to a wide range of consumers (standardized bone china porcelain utilitarian wares for example, appeal to consumers belonging to all socioeconomic classes globally) and they also form a major category of export wares originating from Arita and Jingdezhen and then to European markets historically. However, I am also keen to study the both artists and consumers' counter-reactions to mass-produced utilitarian wares and unique artistic works are also included in this volume as a contrast to the mass-produced pieces.

Brian Moeran's study is centered on a rural community (Sarayama) specifically. There are certain advantages in studying a selected community without comparative elements, e.g., the ability to carry out detailed observations of the community, the process of ceramic product development and community mobilization of resources. At the same time, there are also certain advantages in comparing across locations and communities. Because I am interested in comparative aspects such as comparisons of rural/suburban-based versus urban potters as well as energy use shaped by various priorities for the local environment, I did not restrict myself to one community or location but compared four major ceramics-making locations. Moeran's anthropological publication is based predominantly on fieldwork. As an area study specialist and historian, I have utilized textual analysis, interpretive work, material artifacts examination, library research, social and oral historical methods of corroboration and socioeconomic analysis to complement my fieldwork. The eclectic approach enables me to better analyze the complexities of these locations historically and through a multidisciplinary area studies approach.

Moeran defined folk art aesthetically in his Introduction and Chapter One[2] and conceptualized its broad features primarily as an aesthetic and art movement. In my volume, I have used material artifact identification based on historical ideas of classification and technical details to classify porcelain wares into various categories. For example, classical Arita and Jingdezhen designs and forms are recognizable through their aesthetic appearances (e.g. bone china tablewares, earthenwares, blue and whites and others) and functionalities (e.g. utilitarian kitchenware, industrial products, toilet fixtures, etc). The categorization of aesthetic and functional features are identified and formulated by collectors, historians, archeologists and discerning consumers who aligned those features with ideas about aesthetic appreciation collectively to make up the consumption experience. Moeran's study of *mingei* is also diachronic. After the *mingei* movement matured, folk crafts were defined through form and shape or voluntary self-identification with the folk arts movement. In this volume, I applied a synchronic view of ceramics-making, noting that historicity is important in differentiating between the different styles of porcelain designs as they evolved according to specific historical periods (e.g., Western interest in Arita's exports to make up for shortage of Ming dynasty-era Jingdezhen porcelain supply or contemporary "Arita-style" wares made outside Japan to serve consumers that are price-conscious according to prevailing economic conditions, etc.).

Moeran's insights into internal family structures, social rituals, marriages and community discrimination in Chapters Two and Three[3] are important and provide a glimpse into the daily lives of the ceramicists. I did not probe to the same extent the personal lives of my informants, perhaps this can be a follow-up project in the future. In this project, I was more interested in their views about fuels use and energy choice for firing ceramics. In Chapters Four and Five,[4] Moeran goes into intricate details about the formation and shape of labor specialized in raw materials processing, clay handling, kiln preparation, fuel accumulation and community-based mutual help. These are valuable insights into production factors found within a single ceramics-making community. My approach in this area is based on comparative study and I pay special attention to energy use in the four case studies that I cover. This distinguishes the current volume from Moeran's publication and other existing studies of ceramics industry in at least three ways. Moeran's publication for example studies the Sarayama production ecology and its traditional firing methods using timber. I was not able to find other monographic studies covering different types of kiln technologies and only specialized kiln manuals that I encountered in Japan examined newer kiln technologies that utilize natural gas and electricity. Existing literature do not sufficiently study the significance and impact of energy and environmental factors on ceramics production. In some of the case studies, the production ecology of the pottery-making towns is shaped by environmental factors, such as the presence of an urban infrastructure or ethical concerns about toxicity of production materials and substances.

Chapters Six and Seven[5] of Moeran's book examine environmental and social changes in the ceramics production industry and economic trends that resemble the energy and environmental contents of this volume. In these two areas, the historical approach of this volume adopts a synchronous view in associating technological, economic and social changes with macro changes in the ceramics-making industry. This study centers kiln technology as the most important component of the production process, especially in determining the color, shape and form of the final product. Its output capacity, productivity, environmental-friendliness have direct implications on the final shapes and designs of the wares themselves. Based on the points above, the three main contributions of the volume to existing literature is to: (1) study energy transitions in ceramics product, understand the socio-technological factors behind them; (2) study the comparative histories of the ceramics-making clusters covered in the volume and understand how their urban and suburban make-ups influence ceramics production; (3) understand trends in ceramics designs and production requirements in the context of modernization, consumption post-modernism and gentrification of art districts.

I will now discuss the chapterization of the volume. In Chapter One, I will be examining literature related to the historical dissemination of porcelain-making knowledge from continental Northeast Asia to Japan via trade and cultural exchanges. I then traced the trade routes of Imari (Arita) porcelains from Kyushu to consumers in Southeast Asia, the nearest ready pool of consumers for Japanese Imari porcelains. Secondary literature examines some of the archeological evidence in this aspect. Manila, Tonkin in Annam, Martaban in Burma, Taiwan and Okinawa all became important centers of trade and ports of call. From regional consumers, Arita trade routes started extending further away to Western Europe, particularly during historical periods when Jingdezhen supplies were interrupted by political transitions in Ming to Qing dynasty China. A large volume of this trade was carried out by mercantilist transnational companies like the Dutch East Indies (VOC). In the midst of this exchange, cross-pollination of influences occurred as Arita and Jingdezhen interacted with each other before jointly leaving behind their lasting influences on Dutch Delft and other European porcelain-makers. Modern influence went in the opposite direction as technologies and techniques made possible by the Industrial revolution in Western Europe migrated to Japan which came out of its isolation (*sakoku*) policies. Japan as an early recipient of this modernization process got ahead of other traditional Asian kilns like Jingdezhen to produce sophisticated high-quality bone china Western-style pieces. Japan held on to a technological lead which she continues to maintain today in a highly-competitive industry. Exposure of Japanese export ware pieces to Western customers and collectors in international expositions strengthened and legitimized the global status of Imari (Arita) designs and wares. In Chapter One, therefore, I argued that Arita wares were "re-centered" from peripheral copies of Jingdezhen wares to fulfill European market demand to become a worthy recipient of modernization initiatives incorporated into traditional ceramics-making culture. This process enables Arita wares to become a leading genre of porcelain in its own right.

The implications of a global ceramics-trading network are threefold. First, one historical factor driving and influencing the ceramics industry is consumption. It is a motivator for modernization, technological upgrade and the use of more efficient fuels to generate a larger production output. When consumption grows to a critical size and volume, it can drive new designs, innovation and indigenization of foreign influence and imported styles. Consumption also drives hybridization of design style and forms, combining traditional and modern designs, Eastern and Western influences, cross-pollinating influences from different cultures in the process. In some cases, there is also reverse-osmosis as imitator cultures playing catch-up become even more innovative and attractive to consumers. Sometimes, they even re-influence their original sources of inspiration. Eventually, market-driven consumption determines the viability of designs, enduring sales of a specific category of export wares and the viability of the modernization efforts themselves.

In Chapter Two, still on the theme of modernization, I argued that energy use is an important component of the modernization process. Without improvements made in fuel use, especially in the utilization of more efficient fuels and the advanced kiln technologies needed to fire them, modern mass production of standardized products would not be possible. Therefore, this chapter looked at the evolution of energy use in Arita Japan. In the 1960s, fuel efficiency is counterbalanced by environmental concerns. At the same time, there was another countermovement to return back to older modes of production to rely on timber fuels to produce uniquely-made ceramics (such as folk art or *mingei* pieces) as a counter-reaction against the unmitigated standardization of aesthetic tastes and mechanistic production methods. The emotional factor of nostalgia disrupts the otherwise unproblematic progressive use of fuels based only on productivity. Some ceramicists yearn for a return to the older methods of production and traditional fuel use to recreate old ways of firing porcelain to revive certain designs or remake older designs but with innovative modifications to suit update consumer tastes.

Chapter Three orientates my research work towards a general discussion of modern energy use in the ceramics-making industry in Northeast Asia in general. Are there detectable general trends in energy use in East Asia? As the first country to industrialize in the region, Japan became a disseminator of knowledge related to kiln technologies to the rest of the region starting from the modern Meiji period (1868 onwards). The caloric value of the fuels is first and foremost a determinant of the fuel progression from timber to other fossil fuels like coal and oil. The postwar period starting from the 1960s saw the introduction of the environmental criteria and eventually advanced kiln technologies with automated features and precise temperature controls made one-man-operated (OMO) kilns possible. There are, however, limitations to the effectiveness of the progressive fuels. While newer fuels or advanced technologies may result in lower emissions or greater product output, it cannot resolve some aesthetic requirements, such as avoidance of predictable results of precise firing hampering creative randomness in product output offered by wood-firing. Fuels alone do not determine the viability of production sites. Energy and kiln developments are embedded within a comprehensive production network. In Chapter Four, I point out that Arita successfully modernized and developed a global distribution network, becoming a recognizable brand or genre of ceramics in the industry. In terms of production factors, technologically-driven ceramics-manufacturing relies on a comprehensive production and retail ecology to promote, sell and distribute Arita wares within Kyushu, nationwide and globally. In addition to the use of advanced kiln technologies, design innovation, public awareness of this genre of ceramics and the presence of an entire community with a stake in the success of Arita wares are all necessary ingredients for sustainable production. In terms of global influence, Arita's success is evidenced by the presence of imitative Arita-style wares produced outside Japan.

While productivity, sales and distribution are important, the sustainability of production based on its impact on the environment has also become a growing factor that shapes the practice and technological development of the ceramics manufacturing industry. Given the growing importance of environmental considerations from the 1960s onwards, Chapter Five is a comparative study of kilns in Japan and Hong Kong to examine the level and nature of environmental consciousness in these locations. There are two cognitive levels in which environmental concerns are considered. First, at the level of creative conceptualization, some of the ceramics artists amongst my informants integrated their ideas about "nature" and "environment" into their works, expressing popular concerns and imaginations about the environment through their works of art. Second, aside from the conceptual level, environmental consciousness is integrated into the production stage, e.g., ethical use of environmentally-friendly materials, recycling habits and the use of low or no-emissions fuels. These factors are extremely important in a densely-populated urban location like Hong Kong.

Chapters Five to Eight focuses on the second case study of Hong Kong in this volume. It traces the history of pottery use in Hong Kong. First, it looks at how utilitarian wares produced by traditional home-grown kilns were eventually supplanted by foreign imports due to economic factors—the pricing factor as well as issues related to demand and supply. Imported wares (particularly from mainland China) were more affordable and out-competed domestically-made porcelain industry. At another level of analysis, succeeding generations of Hong Kong potters consecutively created works of arts that represented a hybrid combination of East-West influence integrating the traditional Lingnan styles into their products. To probe deeper into the Hong Kong-based artistic ceramics scene, I conducted fieldwork interviews with the fourth generation of ceramicists in Hong Kong to understand their environmental consciousness, trends in producing works of art, energy use patterns and utilization of advanced kiln technologies to fit the urban environment of Hong Kong.

Aside from the hardware (kiln technologies, fuel, environmentally-friendly ingredients for production), production concepts, thinking on the subject and awareness of post-modernist ideals were influential. Just as technological modernization was a conceptual priority for Japan's ceramics production in the late nineteenth/early twentieth century, counter-reactions to this process and its accompanying mass production concept is raised by some contemporary potters. Chapter Six probes the influence of post-modernism on Hong Kong's potters. It tries to find out how they define the term, examines their thoughts on the complementarities and contradictions of traditional values with innovative ideas in designing or creating ceramics. Subjective post-modernist values develop ideas about ethical production, environmental-friendliness, efficient energy use and artistic production—these are features that are not entirely dependent on profit-orientation. Chapter Seven looks at whether Hong Kong artists who have integrated these values into their artworks and lifestyles are disseminating these values overseas to other communities? In one particular case study in this chapter, I will discuss Hong Kong artists' efforts in collaborating with local Myanmarese/Burmese artists in their capacity as members of non-governmental organizations (NGOs) promoting the conservation of traditional art forms. Ethical concerns, receding traditional Burmese influences, the fate of handicraft production, the process of globalization and its impact in negotiating the contradictions between traditions and modernization as well as the influence of post-modernism are some of the factors discussed in this chapter.

In the locations studied in this volume, creative ceramicists negotiate cooperation and collaboration with the state and non-profit organizations (NPOs) to construct artistic clusters that best manage infrastructural needs, public awareness initiatives, exhibition space and other artistic and production needs. Reflecting these collaborative impulses and pluralistic array of interests, Chapter Eight is a case study of an artist village infrastructure in Hong Kong. As an important artistic and creative cluster in Hong Kong, this chapter examines the gentrification process of the Jockey Club Creative Arts Center (JCCAC). Maximizing space, environmentally-friendly green infrastructure, smooth integration with the local community, freedom in artistic expressions and the range of activities offered by the facilities are discussed here. In the process, the chapter highlights how environmental and infrastructure developments influences the institutionalization of an arts cluster. This chapter will employ a standpoint view of reviewing JCCAC from the perspective of its artists in practice. Chapters Nine to Eleven complete the case studies by looking at the developmental history of kiln-firing in Arita, Jingdezhen and Yingge. The similarities between these three comparatives case studies with Arita form an interesting contrast between specialized pottery-making towns with the urban infrastructure and studio potters of Hong Kong. There are more similarities amongst these three towns than with the Hong Kong case studies. Besides historical legitimacy, closely-knit communities, scale of production, the three pottery-making towns also differ from the Hong Kong urban potter studios in the use of fuels, resource management and heritage conservation in traditional ways of firing. I will integrate these comparative perspectives into the chapters themselves. Finally, the concluding section tabulates the comparative features of the locations covered in this study, looking at their similarities and differences in kiln technologies, creative production, energy use, environmental standards and production ecology.

NOTES

1. Brian Moeran, *Folk Art Potters of Japan: Beyond Anthropology of Aesthetics* (Great Britain: Curzon, 1997).
2. Ibid., 1–4.
3. Ibid., 45–87.
4. Ibid., 86–122.
5. Ibid., 123–55.

Chapter One

Japanese Trade Ceramics in the Premodern Global Trading Space

In this chapter[1], I will discuss a brief historiography and literature review of Arita's premodern position in the global trading regime, contextualized against the Jingdezhen narrative. Premodern global trade was first dominated by overland routes popularly characterized by the Silk Road and its participants were mainly located in the vast Eurasian space of this global trading area. While there are many definitions of the Eurasian trading space that included the so-called Silk Road, some of the broadest definitions include the furthest ends of the pre-modern trading world. For example, Konuralp Ercilasun includes Japan in the broadest definition of the silk route at the farthest East Asian end.[2] There are also differing interpretations of the term "Silk Road" but most interpretations include both the overland as well as the maritime silk route. The maritime route and trading space enjoyed certain advantages that the overland trade routes did not, including non-reliance on oases found in desert routes and caravan animals and their ability to carry higher loads, etc. It also empowered maritime island entities like Japan to participate fully in the global trading system due to the fluidity and reach of the maritime space. The maritime space also facilitated a pre-modern version of intra-regional trading network found in Northeast Asia but extended into Southeast Asia.

Intra-regionally, within Northeast Asia, Hugh Clark quoted Li Zhao in the eighth century who argued that southeastern Chinese districts were all accessible by water and this coastal route served Korea and Japan as well as the intra-Northeast Asian regional trade and, with the expansion of maritime trade during the Song (Sung) dynasty, Japanese and Korean documentation noted the arrival of merchant ships from Chinese port.[3] Song dynasty also saw an expansion of Chinese merchant trade southwards into Southeast Asia.

Other interpretations prefer to conceptualize the maritime and overland route as two parts of a unitary entity with reliance on one when the other met with environmental difficulties. Such views can be found in academic discussions and other settings. Jerry Bentley, for example, noted that sedentary empires like the Tang and Carolingian based their trading systems on trans-Mediterranean-East Asia landed routes and sea-based maritime in the Indian Ocean.[4]

According to John Major, writing in the Asia Society website, the maritime route between Japan and Southeast Asia was sometimes considered as alternative trade routes when environmental and trading conditions were unfavorable overland.[5] In the overland route, Northeast Asian goods traffic appear to move from Northeast China (Shandong and Liaodong) to the Korean Peninsula and then to Japan.[6] During this period, it may be possible that Japan was a recipient of Silk Road goods and comparatively less of an exporter.

Japan's re-centering into the global trading space appears to be associated more with the maritime route, being an island maritime nation, particularly during the Chinese Ming-Qing gap which opened up trading opportunities for Japanese export potters. Therefore, following up on this interpretation, Japan was integrated more comprehensively and fully into the global trading system when the overland caravan route was less practical and when supply from continental economies were comparatively weaker due to internal strife for example. Prior to the seventeenth century, however, given that Japan was located on the far end of the overland Eurasian space, it was geographically distant from the main trading area and often depended on its continental Northeast Asian neighbors for transmission of ideas, technologies, goods and products. Such transmissions could be channeled through the continental route either directly from China or via Korea or through pre-seventeenth century regional maritime trading networks centered around East Asia.

With increasing activity in the global maritime trading space and advancements in navigational equipment and knowledge, however, particularly for Japan after the early European entry into East Asia, the global transition towards the maritime route benefited Japan as it was now less peripheral in terms of reach. It was slowly extending its trading space to Southeast Asia and finally Europe and, in the modern times, the United States within the space of two to three centuries. The global mapping of this re-centering process as Japan moves from the global trading periphery to an active participant may be traceable through industry-specific studies, such as its export industry, specifically the trade ceramics and porcelain industry. The ceramics/porcelain export trade was a microcosm of his Japanese export trade. First emanating from Hirado (in the district of Matsuura) to China and Southeast Asia, it was shifted to Deshima (also spelt Dejima) in 1641 which saw Dutch participation.[7]

Materially, studying the export ware industry has one advantage. Pottery and porcelain have been amongst the most durable artifacts that can survive the ravages of weather and external environmental wear and tear (including seawater) and become archeological time capsules for historical studies. It is relatively more durable when compared against perishable artifacts like textiles, paper, leaves, wood-based material, bamboo, etc. Silk, another major item in the trade routes, for example, is far more perishable than ceramics. Through export trade ceramics, archeologists and historians have been able to investigate the distant past and concluded that, in some cases, they were important for tracing maritime routes and historical trade routes and patterns. The surviving fragments of porcelain shards becomes a potentially important source of information about pre-modern Japan's trade routes to the rest of the world, contributive towards knowledge accumulation of Northeast Asian participation in global trade.

This chapter is interested to see how contemporary scholars within and outside Northeast Asia study center-periphery relations through the idea of cultural transmission by examining discussions on Japanese trade ceramics using the example of Arita/Imari export wares. Ren Seikichi argues that the periphery of the core nucleic Sino-centric universe received cultural influence from China, but it indigenized these ideas in its own ways.[8] Along with the introduction of Westernization and modernization and the influence of empire-building, Ren argues that Japan veered further away from the core-periphery ideas of the Sino-centric world.[9] The idea of a center-periphery system is not static and constantly changes, negotiated by developments and evolution of technological improvements, design innovations, shifting centers of economics and trade, developmental trajectories and the historical sensitivities of cultural trends to these areas. To provide an empirical study, the author adopted the material culture of Arita/Imari trade ceramics to understand the dynamics of Japan's regional and global position in the transition from periphery to the core and center of the global trading system.

There are important debates over the definitions of Imari and Arita ceramics. Five schools of thought are highlighted briefly. The first school of thought attributes Arita and Imari wares geographically to porcelain made near/at Arita (the group of six kilns located at Ichinose, Hirose, Nangawara, Ōhe, Hokao, Kuromuta)[10] and shipped out from the port of Arita. Consequently, chronological and stylistics details are less important. The second school of thought defines Ko-Imari chronologically as export wares made before the 1880s and use stylistic guides to determine this, while those made after the 1880s are classified as Arita wares. The third school of thought relies on stylistics and therefore, even porcelain made imitating Arita or Imari styles but whose kilns are not located near the port of Arita are also classified as Arita or Imari. Other stylistic classifications exist. The Idemitsu Museum of Arts, for example, noted that the Hizen porcelain production facilities were the earliest porcelain makers and their product output could also be known as Imari or Arita porcelain with two particular representative styles of Nabeshima and Kakiemon.[11] The fourth school of thought focuses on the definition of porcelain and locates Arita and Imari as porcelain that developed out of Karatsu proto-porcelain technology and was made from kaolinite clay found in Tengudani by potters traceable back to the pioneering Korean potter who became a naturalized Japanese, Ri Sanpei.

The fifth school of thought is orally and imprecisely circulated amongst the collector communities (such as those in Southeast Asia) which classified *Ko-Imari* as simply old "antique" Imari porcelain wares (usually those made before the late nineteenth century), while Imari can include "antique" porcelain up till near the contemporary period and Arita are basically mass-manufactured new kitchenware porcelain recently available in supermarkets and provision shops in the post-1945 period. This last school of thought, based on antique collectors' classification, is probably the most imprecise and informally defined. In another interpretation circulated within the Japanese collectors' circle, the earliest specimens of Arita porcelain products fired at the start of existence of the kilns are known as "Shoki Imari" (spelt Shoki-Imari in some English-language books).[12] All these ideas about Imari and Arita are important debates but because this chapter is interested in tracing trade routes and understanding cultural dissemination in a re-centered core-periphery format, it adopts the broadest definitions of Ko-Imari, Imari and Arita.

The definition of trade ceramics is comparatively less contentious. Barbara Seyock offered a definition of trade ceramics consisting of celadon, porcelain and stoneware from East Asia that was exported through trade routes, particularly, in Seyock's view, the maritime trade route.[13] Much of these trade ceramics came from China, widely regarded as a major ceramics-producing center for the world in many world historical periods and recognized for its material use, production techniques, artistry and innovations. And other originators of trade ceramics were peripheral to this ceramics-producing center before the late pre-modern period.

JAPAN'S EARLY TRADE CERAMICS

Even before the late pre-modern period, however, Japan's trade ceramics did play a minor role in global export ware trade before the seventeenth century. The Tokyo National Museum pointed out that, according to some of the oldest records related to Imari, in 1647, Imari export wares were shipped to Cambodia through Chinese vessels.[14] Besides the Chinese, early European arrivals were also amongst the earliest to trade in Imari export wares. According to Sasaki Tatsuo, after Spain annexed the Philippines in 1571, a limited volume of Japanese pottery (along with larger quantities of Chinese ceramics) were exported to America by Spanish ships departing from Manila and this marked the beginning of the maritime route for the global ceramics trade.[15]

Sasaki based this Pacific trade route argument upon the discovery of limited quantities of Japanese ceramics (mixed with Chinese ceramics) in the Acapulco beaches as early evidence of Japan-US trade that depended on two middlemen, the Spanish (from East Asia to America) and the Chinese (junk transportation from Japan to Manila).[16] Sasaki's argument are cutting-edge but also potentially controversial as he argues that European ships created a "ceramic road" that reached both the Pacific and Atlantic Ocean, extending to as far as Mexico and Cuba where both Chinese and Japanese ceramics excavated shards could be found and some of them showcased in Cuban museums.[17] Whether one agrees or disagrees with this interpretation, in terms of quantity, this was a small start to a process that would take nearly two centuries for Japanese ceramics to become globally traceable, particularly in terms of export ceramics manufactured and shipped worldwide by the Japanese themselves without the need for middlemen.

Increasing amount of scholarship indicates that formerly peripheral ceramics-producing areas had also experienced their own trade historical trajectories in terms of ceramic wares exportation. In this essay, Japan's empirical example is studied, not as a center-periphery study in traditional Sino-centric terms, but as another perspective on mapping trade exportation out of Northeast Asia in the pre-modern period highlighting the dynamic and fluid nature of moving centers of global trade, particularly on a global scale. Perhaps, two of the earliest pre-modern continental sources of influence on Japan probably came from Korea and China. According to Francois-Bernard Hyghe, the ancient territory of Silla in Korea was once a connecting point between China and Japan and, during the seventh to ninth centuries, a strong and unified Silla coincided with a golden age in China (the Tang dynasty) as well as a centripetally-governed Japan (central control over clans)[18] that created stable environmental conditions for a Northeast Asian platform of trade and exchanges. Japan's reception of goods from the Silk Road trade may be exemplified by the iconic Shōsō-in museum of treasures at the city of Nara where gifts and trade goods from the Silk Road were pristinely conserved for posterity.[19] Such continental arrivals of influence and exchanges would eventually trigger off Japan's indigenous development of a trade/export ceramics industry through the influx of materials, technology and eventually human talents.

A mainstream interpretation of the origins of ceramics development in Japan focuses on the introduction of Korean potters into Japan after the latter's incursions into its neighboring state between 1592 and 1598.[20] According to Nancy Schiffer, Ri Sampei (or Sanpei, one of the Korean porcelain makers transported to Japan by Hideyoshi) first uncovered porcelain clay in Izumiyama in Arita Kyushu to feed a production system that churned out export wares for the Dutch who introduced it to the European markets.[21] Nagatake Takeshi's seminal work on Imari and Kakiemon porcelain, however, provides an alternative to this theory, arguing that ceramics-making may have started in 1605 which is earlier by a decade led by Ienaga Shōemon, but this group was displaced in Tengudani by the arrival of Korean ceramics-makers from Taku.[22] Nagatake's example of the Shōemon group suggests that Japan's indigenous ceramics/porcelain manufacturing capabilities and history may be more dispersed than centralized, that there may be several (or at least more than one group involved in the process) groups pursuing ceramics/porcelain-making from different times and spaces.

Even before this influx of foreign-born potters, Japan already had physical samples of export wares which they probably purchased through trade from Korea. Morimoto Asako's important archeological study of 120 excavations in Hakata, carried out since the initiation of the first subway construction in that area, is useful in detecting the influence of Korean export wares on Japanese trade ceramics development throughout the fifteenth century before the arrival of Korean potters to Japan in the 1590s.[23] In other words, the Korean potters arrived in a location where there was already an established tradition of admiring, emulating and study of Korean pottery.

EUROPEAN INTEREST

With the growth in Japan's indigenous improvements and upgrades of the techniques and technologies that she had absorbed from Korean potters and also manufacturing development based on import substitution of quality ceramics from the sixteenth to eighteenth centuries, Japan gradually caught up with Korea by absorbing Dutch-originated European knowledge and progress made in indigenous technological improvements during periods when Korea was less active with foreign trade and exchanges.[24] According to Schiffer, Japan exported porcelains (mainly blue under-glazed wares) from 1650 to 1700 to Europe, South Asia and Persia.[25] At the time of 1700, Japanese as well as China's porcelain were transacted and transported to London and Amsterdam, which were themselves important cities of commerce and main locations where Asian porcelains (including Japanese ones as a main component) were redecorated by European craftsman.[26]

While the Europeans were redecorating Japanese wares, back in Asia, Japan was able to study and acquire Dutch learning and knowledge partly through trade which included Japanese porcelains amongst its items. The Dutch merchants and the Vereenigde Oost Indische Compagnie (VOC)[27] exchanged Chinese silk, European textiles, Indonesia spices, Thai and Taiwanese animal skins, as well as African and Southeast ivory for Japanese porcelain, precious metals, lacquer ware, gains and camphor.[28] The caption for the "Hasami ware (Japan): Dish with incised waves and lotus design" displayed at the Art Gallery of New South Wales (NSW) noted that the Dutch shipped Japanese porcelain wares to Indonesia's Sulawesi in the 1640s and 1650s, although celadon dishes from Omura located next to Arita was more common than Arita porcelains.[29]

Introduction of foreign expertise and technologies from Korea alone may not immediately translate to overseas and local consumer demand necessary to jump-start new industries. Foreign competition, particularly high quality export ceramics from China, depressed demand for indigenously-produced ceramics until China's transition from Ming to Qing dynasties, reduced supply of Chinese ceramics imports from the middle of the seventeenth century and the resulting gap in supply created the demand for domestic production of Japanese ceramics.[30] This was an important stimulus factor in jump-starting Japan's export ware industry on a sizable scale. Another interesting factor, cited by Nagatake, aiding Japanese exports was the fact that in the seventeenth century, consumption trends in Europe had moved from precious metals to glass and ceramics, often with the latter denoting class and prestige for the European imperial and royal nations like the Dutch, the British, the Germans and the French.[31] In fact, in February 2011, the UK Department for Culture, Media and Sport (DCMS) blocked the sale and export of Spencer family heirlooms, the early-eighteenth century Imari porcelain, which reflected "manufacturing techniques, European taste and trade between Japan and Britain in the late seventeenth and early eighteenth centuries."[32]

One of Japan's main external customers during this period were the Dutch, who made handsome profits in trading in Japanese export porcelain that made up for the drop in Chinese supply. While the Dutch rapidly imported 200,000 pieces of Chinese porcelain in 1644, the conflicts in the Ming-Qing transition prevented further increase and eventually, from 1654 to 1684, the supply dried up.[33] The shortage in Chinese porcelain caused prices of porcelain wares to go up in Holland, as Christine Moll-Murata points out, during the Ming-Qing transition, when Japanese Arita substitutes were the available alternatives, the cost of a porcelain plate or bowl was four to nine times more than a majolica equivalent for example.[34] By the mid-seventeenth century, Japanese production system had become more sophisticated and specialized as production sites saw its minders hiring laborers trained and skilled in throwing, decorating and modeling.[35]

Schiffer noted that the Dutch established a porcelain production site (VOC) in Hirado (later Deshima) which managed blue and white export wares resembling Kraaks and Chinese Imaris not available in China due to the Ming-Qing transition and so from 1653–1682, Japanese export ceramics became substitutes in this area.[36] Initially emulating Ming designs and features, Nagatake pointed out that the Ko Imari porcelain indigenized and became more uniquely Japanese, with a variety of influences from Muromachi ink paintings, Kanō and Tosa screen art, popular culture and lacquer art (*maki-e*) during that period.[37] Nevertheless, the shortage in supply continued to drive the demand. Due to proximity, Indo-China became a natural market for Japanese porcelain (for example, Hizen porcelain originating from Nagasaki in the 1600s).[38] This was near the time when the Vietnamese displaced Chinese as the main silk supplier and a Japanese trading node was set up in Da Nang in the late sixteenth century which operated till the Tokugawa policy on seclusion in 1635.[39]

The Japanese were also active in other parts of Southeast Asia too. Nogami Takenori's article noted that, from late 1640s, export wares were exported from Nagasaki to Cambodia transiting through Thailand.[40] During the Ming-Qing transition's reduction in Chinese export pottery, Hizen shards and artifacts can be found further beyond Indo-China to the Malay Peninsula and the Indonesian archipelago, while Nogami's own groundbreaking research in 2004 extended the perimeter of Hizen ware exports to the Philippines.[41] Thi Ha Thanh Nguyen's study provides an interesting glimpse into this trade from the Vietnamese perspective. Between 1600-1800, when Vietnam was divided into two kingdoms, including the northern Kingdom (Tonkin), which was ruled by the Lê dynasty and Trịnh Lords.[42] Lord Trinh emulated the success of his rivals in the South in trading with the Europeans and invited the Dutch to participate in North Vietnam's trade and the Dutch quickly spotted the market need for trade ceramics caused by the dip in Chinese export wares supply due to the *hai jin* law in the Ming-Qing transition and became a distributor of Japanese porcelain wares in Vietnam.[43] In this sense, the Dutch became an arbitrator of Japanese porcelain products between Japan and Southeast Asian consumers.

The geographical area of consumer markets for Japanese export ceramics gradually expanded from nearby (Southeast Asia) to non-Asian locations (for example, Western Europe). Schiffer argues that the Industrial Revolution generated surplus productive resources like time and purchasing power, which allowed European consumers to spend more on higher grade consumer items like intricately-designed porcelain export wares.[44] Because of the drop in Chinese porcelain supplies, Japanese ceramics makers were quick to pick up Chinese ceramics art stylistics and sold Chinese-style porcelain to the Dutch although the numbers started off small (according to Moll-Murata: 11,500 in 1611, 48,000 in 1663, 65,000 in 1665) because of the higher prices of Japanese export wares (two to three times higher).[45] Moll-Murata pointed out an important caveat, however, that Christiaan Jorg noted unofficial trade in Japanese porcelain meant that the numbers were much higher than what was officially recorded.[46]

Some of the Korean potters and artists who found themselves settled in Arita Japan in the early seventeenth century started decorating porcelain in the commonly-favored colors of blue, gold and red with cobalt that is coated with a layer of glaze before they were exported through the Dutch middlemen who were waiting in the ships berthed at Nagasaki harbor.[47] Two genres of Japanese export porcelains appear to stand out during this period—the blue and white as well as the polychromes. Besides blue and white porcelain, the Ming-Qing transition supply gap created consumer demand for Imari polychrome wares (the gold and red items mentioned above) in Europe.[48] From a Chinese perspective, the quality standard and results of Imari production at Arita was not ideal between 1630s and 1640s, but the pressure of high orders and mass manufacturing eventually made the quality comparable to those found in China.[49] If this was the case, it may have been reflected in the large orders from Europe. In 1659, for example, the East India Company of Holland (VOC)[50] requested for 56,700 pieces of ceramics to be made at Hizen and such acquisitions, Sasaki argues, eventually influenced Japanese ceramics design and manufacture.[51]

Archeological and textual evidence both appear to corroborate and confirm this period of high Imari export ware activity in Europe. According to Oliver Impey, the oldest shard of Japanese ware (probably Shoki-Imari) in the West was discovered in Amsterdam in 1973 attributable to the early 1650s, and which may have been from a batch of cargo to a commercial site.[52] According to Sakuraba Miki, the first officially-recorded trade of porcelain goods exported from Japan to Holland in 1659 consisted of tea cups and thereafter, about 10,000 of them went from Japan to Holland annually.[53] Sakuraba has found Japanese *Ko-Imari* porcelain cups that were hand-painted and accompanied by saucers datable back to the 1670s to 1690s and, in terms of official documentation, Dutch East India Company (VOC) referred to them as *thee pierings* or tea saucers from 1659.[54] For an example of the Japanese trade wares that were exported from Japan to the West during this period, a late seventeenth-century big Imari hard-paste plate made in the port of Arita in Kyushu decorated with underglazed cobalt blue and overglazed enamels in red and gold colors can be found in the Seattle Art Museum and, according to the Museum, they were once used to adorn the walls of European porcelain rooms as displays.[55]

INFLUENCING (SOME) CHINESE CERAMICS

John Ayers, Oliver Impey and J. V. G. Mallet argue that because Japanese ceramics, in particular the uniquely designed and drawn ones in the 1670s, had grown so popular within Europe (including the Dutch) that the Europeans started purchasing them from southern Chinese sources, as evidenced by the collections found in the UK, Germany and France in addition to the Dutch.[56] This argument is taken further by other scholars who attributed

Chinese stylistics to emulating Japanese templates. After the Ming-Qing transitional period, Chu Lung-hsing's important essay makes the innovative argument of the important influence of Japanese ceramics images of the Dutch people on Chinese porcelain. Chu argues that Chinese representations of Westerners, such as those like *Huangqing zhigong tu* (illustrative text) found in the sixteenth year of the Qianlong Qing dynastic rule (1711–1799), derived their pictorial understanding of the Dutch people from Japanese representations of the Dutch detectable through, for example, the poses struck by the figures in the pictures.[57]

Chu Lung-hsing speculates that some of these Japanese pictorial representations may have been in the form of Edo prints brought over by VOC or the Dutch East India Company from Japan to China while others may be in the form of porcelain wares.[58] Chu cites a pair of Chinese ceramic bottles dating back to the Kangxi Qing dynasty era (1654–1722) that displayed pictorial representations of the Dutch people which bears certain resemblance to Japanese Edo prints.[59] This creates an interesting situation in which Northeast Asian design centers started influencing each other in cultural representations of Westerners and the West. Perhaps, as equally interesting as the influence of Japanese designs on Chinese portrayals of Europeans is the influence of Japan on Western depictions of East Asians. Chu cited the example of Amsterdam-based ceramics artist Cornick's adaptation of a Japanese plate design which replaced the featured Japanese (and Chinese) women under an umbrella in the original design with a Dutch individual.[60] This is a curious case of a Western cultural adaptation of a Japanese image based on a Western-inspired design seen through the gaze of the Japanese.

More importantly, by influencing both Chinese portrayals of Westerners and Western depictions of East Asian, Japan had effectively become an arbitrator of East-West interactions and an intermediary of fusion styles, a role that would be highly accentuated in the modern era and perpetuated and brought to new heights in the contemporary period. Positive Western reception to Japanese cultural products was detectable in the early modern period in mid-nineteenth century when artists like Vincent Van Gogh picked up Japanese popular cultural products and design items.[61]

Japan's historical role in East-West fusion of styles and images in export porcelain design was now transformed into something much larger and broader in scope covering a wide expanse of popular cultural items and the identity of its designs dating back to this East-West exchange appear to inspire a unique identity that has endured through the times until the contemporary era and, at the same time, remained as an object of emulation both in the West and in the East, as Mackay in the CNN news report appear to suggest.

JAPAN AS CONSUMER

Like the case of permeable cultural influences, ceramics trade between Japan and the rest of the world was certainly not unidirectional. Japan represented a major consumer market for foreign porcelains in many historical time periods. From near to far, evidence of porcelains from other world regions to Japan has been consistently found in Japanese excavations. In 2005, Sakai Takashi produced a report which described the discovery of Turkish Iznik enamel shards in Japan which represented the golden period of the Ottoman Empire dating back to the early Edo period (seventeenth century) in the location of the Maeda Residence.[62] John Stevenson and John Guy noted that in the sixteenth and seventeenth centuries, artefacts in *daimyō* (feudal lords) and traders' family possessions passed down from generations to generations indicated that Vietnamese ceramics was in fashionable use by practitioners of the Japanese tea ceremony (*chanoyu*).[63]

Historically, Japan was certainly a major consuming market for Chinese ceramics. According to Evelyn and Thomas Rawski, before the European influx of Chinese porcelain from the sixteenth to the eighteenth centuries, global distribution of Chinese porcelain and stoneware (white wares, *qingbaci* [white with bluish tinge] and blue and white porcelains or *qinghuaci*) were distributed by Chinese, Southeast Asians and Arabs to the major consumer markets of Japan, Korea, Southeast Asia and the Middle East.[64] This might be an earlier form of pre-modern maritime trade of export porcelains to Japan which saw Japan in the receptor role as an absorber of cosmopolitan goods. According to Evelyn and Thomas Rawski, Japanese documentation indicated the importation of 371,000 Chinese porcelain items annually between the years 1635 to 1645.[65] Sub-regionally within China, Japan was also an important consumer market for Southern Chinese porcelain and traders/merchants. Lin Renchuan highlighted a 1609 private trade example whereby a Chinese group consisting of Lin Jing (from Fujing), Captain Wang Hou (from Zhangluo), Zheng Song and Wang Yi (helmsmen), Zheng Ji, Lin Zheng and others (sailors), Li Ming (pilot), and Chen Hua (Japanese translator) led a group of traders to Japan and sold products that included porcelain to the Japanese in exchange for Japanese silver that was melted in the ship's inbuilt oven and bellows by the on-board silversmiths, Jin Shishan and Huang Zhengxian.[66]

Besides direct sales like the above example, in the private trade between Fujian and Japan, there were also participating intermediaries such as the Taiwanese. The Dutch East India Company utilized Taiwan to access the Chinese market with the aim of setting up a Japan (Nagasaki)—Indonesia (Batavia)—China (Southern China) trade triangle to keep out European competition, particularly from the early arrivals like the Portuguese and the Spaniards.[67] When Taiwan ousted the Dutch, Zheng Jing who was the offspring of

Zheng Chenggong, sold pottery from continental China to the Japanese amongst other consumer markets in exchange for metals and other necessities for the Taiwanese community and economy, as well as the perpetuation of the Zheng clan's power.[68]

Like Taiwan, Manila also became an important intermediary center of trade for Japanese ceramics buyers and pottery acquisitions. In the late sixteenth century, in avoidance of the monsoon storms, Japanese traders from Southern Japan travelled to Manila in the months of October and March/May annually to procure amongst other things, Chinese Tang/Song dynasty-era earthenware buried under the mud of the waters off northern Philippine coasts, as well as other ceramics and porcelains, in exchange for the silks, cotton, wool, copper, gold, iron, tin and pottery goods that they offer.[69] According to John Stevenson and John Guy, Japanese consumers liked tea jars made in Vietnam and traded in the Philippines with Japanese sea-going vessels' trading and business activities in Manila, attaining new heights in the volume of bilateral trade between 1615 and 1625.[70]

Besides Manila, Japan was also a trading nation involved in the trade of Martaban jars from other Southeast Asian sources that were sold commercially in the maritime space between the Indian Ocean and East Asian maritime waters between fifteenth to eighteenth centuries. Some of these ceramics that originated from Myanmar were uncovered in Japan.[71] Nan Kyi Kyi Khai studied a Myanmar-originated white glazed dish and other white wares uncovered in Japan (Osaka and Hirado) as well as black-glazed Myanmarese Martaban jars excavated in Japan (Ōita, Fukuoka, Nagasaki and Okinawa) that originated from Martaban or Mottama port.[72] According to Nan, intermediaries were involved in this trade as Myanmarese wares travelled from the Martaban harbor to possibly ports in Thailand, Philippines and Indonesia before arriving in Japan.[73] Besides Manila and Burmese interests, Tonkin (Vietnam) at times was a trading center for Japanese ceramics in the seventeenth century, but also Vietnamese pottery for Japanese consumption in the latter half of the same century as a surviving record dated 1678 indicated that a thousand rice bowls were shipped to meet Japanese demand.[74]

Archeological excavations in Japan have continued to yield evidence of trade cosmopolitanism in terms of foreign export ceramics trade to Japan. The archeological dig at Ōtomofunai-machi location in Ōita Prefecture by the Ōita city Board of Education between 1996 and 1997 turned out trade pottery material artifacts from Southeast Asian countries like Vietnam, Thailand Korea as well as Northeast Asian-originated shards from China and Korea of various vintage.[75] While porcelain trade was bilateral in nature as foreign export ceramics entered Japan and Japanese export ceramics left Japan for other destinations in different quantities during different time periods, the process also made Japan both the receptor and innovator of design and artistic traditions and influences. As bilateral trade took place between

Japan and the rest of the world, Japan started establishing itself as an originator of indigenous design stylistics and techniques.

Sasaki argues that Japanese and Chinese ceramics had influenced their European counterparts in the vessel shape, design and form through the global maritime trade in ceramics.[76] The volume *Tōji no tōzai kōryū ten: Arita, Derufuto, Chūgoku no sōgo eikyō* is interesting in the sense that it presents materials evidence visually to indicate what the book argues is the mutual influencing of each other's culture between the Dutch Delft, Japanese Arita and Chinese porcelain through a comparison of the porcelains from these three sources although detailed textual information on this subject matter in the volume is light.[77]

Other scholars also try to look for evidence within the area of stylistics. Moll-Murata pointed out in her essay that, between 1680 and 1725, the Dutch Delft ceramic manufacturing system began to emulate Japanese decorative techniques and incorporate them into the technical and aesthetic aspects of production.[78] Moll-Murata attributed the emulation of Japanese Imari stylistics in trade ceramics to the widowed business owner in Het Moriaenshooft, Madam Jannetje van Straeten, and her ceramics series De Grieksche A (the Greek A factory).[79] This marks the gradual centering of Japan within the trade ceramics industry in the production and aesthetics aspect. It would take another 150 years for Japan to acquire a leading position in the design aspect of this industry. Japanese adaptations of Western styles in export porcelain and Western emulation of these Japanese adaptations eventually became a major and important force of design influence by itself with its own consumer following and process of continual innovation. Training centers in manufacturing sites sanctioned by artisan initiatives and the private sector contributed to quality enhancements and design standards in the 1880s.[80]

The renaissance in porcelain trade and innovation between Northeast Asia and Western Europe did experience a period of political disequilibrium and disturbance. While the Dutch East India Company (VOC) profited handsomely from the Japanese and Chinese (via Taiwan trade), its revenues were cut when a politically re-energized early Chinese Qing dynasty government ousted the Dutch from Fort Zeelandia and Japan's ruler, the Shōgun, restricted Dutch trade as part of its isolation (*sakoku*) policy.[81] Chinese reassertion of the export porcelain trade was strong especially taken in the context of Japan's isolation policy which closed it to the rest of the world with the exception of only the Dutch and the Chinese traders. Geoffrey A Godden dramatized the reemergence of Chinese export ware presence with the consolidation of the Manchurian Qing dynasty at the end of the eighteenth century when Europe had more imported porcelain from China than those made by the Europe-based porcelain makers taken together.[82]

MODERNIZATION

This situation of China's re-emerged competitiveness with Japanese export wares and superiority in figures remained until America's Commodore Perry arrived in Japan and, with a show of modernized might, opened Japan up to the rest of the world and prompted the modernization of Meiji Japan in 1868, re-introducing modern techniques, science and management knowhow in addition to Western cultural influence to Japan, a process that would eventually globalize Japan's ceramics trade in the modern world. The combination of absorption of Western learning, constant quest for productivity improvements, and knowledge trawled back from the West by Japanese who spent time overseas and understood aspects of Japanese art that Westerners admire enabled Japan to position its export ceramic industry as a top-ranking entity in the world.[83] Japan also modernized its distribution and retail supply chain management. In this early modern or late pre-modern period, the Japanese ceramics traders carried out economic exchanges and trade with Southeast Asia and Northern Europe like Great Britain through what Nagatake coins as a "wholesaling system," which involved distributional stakeholders like the Hizen wholesaler based in Edo, Saga Fief Merchants Association in Nagasaki, Tashiro Monzaemon of Honkōbira, Hisatomi Yojibē of Nakanohara (with the auspices of the Nabeshima fief authorities) whereby these dealers could directly retail to local shops as well as foreign customers.[84]

After Japan rejoined the modern world trading community, evidence of Japanese export wares dating back to Japan's early modernity could once again be found. Nan's paper which studied trade ceramics in Myanmar noted that imported trade ceramics shards were uncovered at the Mandalay division of Bagan in Myanmar and they included a shard of Japanese Hizen ware datable back to the modern Meiji and Taisho periods (1868–1912), something rarely found in the vicinity.[85] The shard, along with others, was determined by Nan to be evidence of maritime trade that found its way to the bank of Twante canal, which was utilized as a trading center from fourteenth to early twentieth centuries accounting for Japanese, Chinese and European wares found here and the Japanese shard was dated through comparison with other similar samples unearthed in Southeast Asia (for example, at the site of the Wolio castle excavation location of Indonesia), UAE and Japan.[86] According to Nan, Myanmarese trade in Japanese export ceramics boomed economically between eighteenth to early twentieth centuries in the Twante port[87] and the artifact served as a reminder of this.

CONCLUSION

By the latter half of the nineteenth century, Japan's ceramics were re-centered from the periphery and became a defining style, design movement and trend known as *Japonisme.*[88] The internationalization of Japanese trade ceramics and stylistics was aided by international treaties, for example, after the Treaty of Commerce between Japan and France was signed in 1858, more goods travelled by the maritime route from Japan to French consumers than before and, from the latter half of the nineteenth century, much of this trade was conducted by Japanese, instead of going through intermediaries or through Western merchants.[89]

In 1862, London's International Exhibition showcased Japanese-style aesthetics to Europe and the 1876 Philadelphia Centennial Exhibition saw the Japanese booth as the most eye-catching space visited by millions.[90] Japanese ceramics were also showcased in the international exhibitions of Paris, Vienna and Philadelphia and sold to avid Western collectors while the modern Meiji government created exposure for Japanese ceramics by organizing local exhibitions for information dissemination and awareness purposes (for example, *Naikoku Kangyō Hakurankai Jimukyoku*, Exhibition Bureau for domestic industry[91] and *Kyōshinkai*) under the umbrella of the *shokusan kōgyō* industrial policy.[92] This was perhaps an early modern example of Japan's famous state-led system of development and industrialization.

As Japanese-style ceramics gained popularity and a foothold in the Western markets, Japan started producing Western-style export ceramics by early twentieth century. Reminiscence of the Iwakura mission, Japan dispatched talents to the West conditional upon bringing back skills learned and information gathered for the domestic audience, and it also invited Western experts to come to Japan.[93] The influx of Western experts was important in two areas: providing new technologies (including technologies that used more efficient fuels) and introducing Western stylistics into ceramics manufacturing (see table 1.1).

In terms of technological progress, Japanese ceramics manufacturing facilities and industries started off with using traditional wood-fired technologies to produce their porcelain. In classifying what is traditional and innovative, it may be important to understand that the evolution of kiln technologies and fuel use is relative in nature. While timber-fired kilns are inefficient and difficult to control compared to oil, gas and electric ones, the timber-fired Chinese dragon kilns were, in fact, considered hi-tech in their time, and Lisa Rotondo-McCord et al. argue that they can keep a "high, accurate and evenly distribute temperature" better than older kiln designs.[94] Prior to using dragon kilns, Japanese kilns were typically minor cottage industries in the form of *noborigama* or climbing kilns located and built into the slopes of hills that could manipulate wind inputs and operate at various heat intensities for mak-

年度Year				
1896	製造戶數 Manufacturing facilities	本窯 Kilns	同上間數	錦窯 Kilns
	34 戶facilities	18基sites	168基sites	100基sites
1931	飯食器類 Dining ware	美術工藝品 Artistic products	工業用品 Industrial wares	電器礙子 Electrical-related items
	142萬円 1,420,000 yen	67萬円 670,000 yen	5.6萬円 56,000 yen	17萬円 170,000 yen
	玩具 Toys	其他 Others	總數 Total	
	1.5万円 15,000 yen	6.2万円 62,000 yen	239万円 2,390,000 yen	
	日常用品/餐具 Utilitarian wares Kitchen wares	工業品 Industrial wares	美術/展示品 Artistic wares	其他 Others
1946	1,100萬円 11,000,000 yen	819萬円 8,190,000 yen	219萬円 2,190,000 yen	149萬円 1,490,000 yen
1947	1,495萬円 14,950,000 yen	1,328萬円 13,280,000 yen	527萬円 5,270,000 yen	210萬円 2,100,000 yen
1946-7	85棟工房 85 workshops	15棟工房 15 workshops	23棟工房 23 workshops	5棟工房 5 workshops
1958	總數 Total	電器用品 Electronics-related products	工業用品 Industrial products	生活用食器 Utilitarian wares/kitchen wares
總值（美金） Total (USD)	18,776	4,014	6,024	4,939
%	100	21・4	32.2	26.3
1958	玩具 Toys	瓦片 Tiles	衛生用品 Health products	其他 Others
總值（美金） Total (USD)	231	2,291	30	1,229
%	1.2	12.2	0.2	6.5

Table 1.1. Production Figures for Arita Export Porcelain. Source: Nagatake Takeshi, *Arita yakimono no dokuhon* [Publication on Arita Pottery] (Japan: Aritatōji Bijutsukan, 1961), 41, 52, 60, 63.

ing porcelain.[95] According to Koyama Fujio, there may be 136 of such facilities in Ri Sanpei's time at the start of Arita porcelain-making history.[96]

But German chemist Gottfried Wagner (1831–1892) transformed the industry when he introduced coal-fired technologies to the ceramics industry, and revolutionized the industry so much that he became known as the creator of the modern ceramics industry in Japan, according to some interpretations.[97] According to Yao Chun's evaluation, Japan's lead in porcelain is due to continuing innovation in new types of kiln technologies, a feature which

Japan has pursued since the start of modernity.[98] The next stage of modernity is signified by the energy transition from the use of timber-fired dragon kiln to coal-fired kilns.[99] This Arita energy transition forms the foundation of the modern concepts of productivity increases. From 1897 to 1906, the Japanese state aided ceramics-making regions with the transition to coal-fired kilns, including the first such contraption built in the Seto Ceramics School in 1902.[100] Coal-fired kilns revolutionized the ceramics industry as previously small scale industries making folk wares were now able to produce highly standardized Western-style dinner table ware as well as other industrial ceramic applications such as construction tiles.[101]

This energy transition came at an opportune time because the Museum of Contemporary Ceramic Art, the Shigaraki Ceramic Cultural Park argued, in their edited volume *Japonisme for the Western World: Pottery of the Meiji Era* that the fall of the Bakufu, the late Tokugawa clan system and the modernization process of Meiji Restoration brought about social dislocations where porcelain and ceramics artisans were compelled to be financially autonomous and therefore making utilitarian wares that were highly usable to feed a consumer market by appealing to their tastes.[102] This coincided in a timely fashion with the advent of Western consumer markets and stylistics influence.

Yao's argument is that continuing innovation in Japanese kiln technologies was the strength of its ceramics manufacturing system, which persisted into the contemporary period. In the process of continuing Westernization, Japan was able to adapt German technologies for their own use and embarked on a natural gas revolution in porcelain manufacturing by the 1970s.[103] If the characterization for coal is affordability and abundance, then the discourse on gas energy is one of long-term commitment and supply, with the imagery of a cleaner form of fossil fuel for future utilization and development.[104] Coal appears to be inefficient, polluting and dangerous compared to cleaner fuels like oil and gas, although gas has contemporarily displaced oil as an even cleaner source.[105] According to Yamashita Eiichi, the ceramic manufacturing town of Seto used so much coal for firing porcelain that even sparrows flying over the town turned black from the soot pollution.[106]

Besides learning and absorbing Western technologies in the use of energy, Japan's exchanges with the West also facilitated a better understanding of its consumer markets in the West to cater to their needs. According to Itani Yoshie, many such export ceramics originated from Japan's Seto region and were exported to a new large market across the Pacific to the United States of America.[107] Early modern trading companies specializing in high quality crafts soon sprouted up like the Kiritsu Kōshō Kaisha (Kiritsu Industrial and Commercial Company) which was an example of early modern state-led and supported firm which had an influential say over stylistic trends, distribution and also management of the inventory of Japanese product outputs from world fairs.[108] They became important intermediaries in dealing with Western firms and consumers. In the West, Japanese export wares were soon available in American companies like AA Vantine & Co. Inc. located in New York city in Fifth Avenue and thirty-ninth Street that sold Imari, Banko, Sumida and other Japanese porcelain export wares (both decorative and utilitarian) to American consumers.[109] Descriptions of Japanese tea sets started appearing in American mail order catalogues, for example, the 1927 Sears & Roebuck Catalogue contained items like a "23 Piece China Tea Set" with the product description: "The decoration consists of a blue lustre band with a black line"; or a August 1929 Butler Brothers catalogue containing the itemized description of "Jap China 23 Pc. Tea Set" with "Double lustre 3 decorations (tan with blue bands, blue with tan bands, and iris with mother-of-pearl bands). Black inner line."[110]

By this time, the re-centering of the Japanese export ceramics is completed as it not only exported, exhibited, showcased and marketed traditional Japanese export wares globally but also produced Western-style wares for export to destinations that originated those cultural styles. Through the ceramics export trade, this Japanese industry was no longer in the periphery of the global trade but very much in the center of a modern globalized international trade network that corresponded with Japan's emergence into modernity. Interestingly, growing confidence of its capabilities and industrial output and quality may have persuaded the Japanese to store their most valuable works (including copies of those produced for international expositions) in their own country in what would become the Tokyo National Museum (Tokyo Kokuritsu Hakubutsukan) in Ueno.[111]

NOTES

1. Part of this chapter on Izumi case study is funded by the CUHK Direct Grant.
2. Konuralp Ercilasun, "Silk Road as a Sub-global Region: A Sphere Emerging from the Interaction of Cultural and Economic Fields" (paper presented at an International Seminar on "Reviving the Silk Route: New Initiatives and Engagements for the 21st Century," International Centre, Goa, Dona Paula, India, February 9–10, 2007), http://www.centralasia-southcaucasus.com/docs/Silk%20Route/Paper_Konulrap_Ercilasun.doc.
3. Hugh Clark, "The Religious Culture of Southern Fujian, 750–1450: Preliminary Reflections on Contacts across a Maritime Frontier," *Asia Major* 19, no. 1/2 (2006): 220.
4. Jerry H. Bentley, "Hemispheric Integration, 500-1500 C.E.," *Journal of World History* 9, no. 2 (Fall 1998): 240.
5. John Major, "Geographical Setting of the Silk Roads," Asia Society, August 18, 2008, http://asiasociety.org/countries-history/trade-exchange/geographical-setting-silk-roads?page=0%2C3.
6. Ibid.
7. Takeshi Nagatake, *Classic Japanese Porcelain: Imari and Kakiemon* (Tokyo, New York and London: Kodansha International, 2003), 52–53.
8. Ren Seikichi, "Higashi Ajia no bunka kankyō no keisei" [Cultural environmental formation in East Asia], in *Kankyō to bunka – (Bunka kankyō) no shosō* [Environment and culture—

aspects of cultural environment], ed. Nagasaki Daigaku Bunka Kankyō Kenkyūkai (Kyushu: Kyushu University Press, 2000), 285.

9. Ibid., 292.

10. Irene Stitt, *Japanese Ceramics of the Last 100 Years* (New York: Crown Publishers, 1974), 40.

11. Idemitsu Museum of Arts, *Kakiemon to Nagashima* [Kakiemon and Nabeshima] (Tokyo: Idemitsu Bijutsukan, 2008), 5.

12. Nagatake Takeshi, *Nihon no tōjiki: Sono kamaba o tazunete* [Japanese ceramics: Visit to the Kilns] (Tokyo: Shakai Shisōsha, 1966), 203–4.

13. Barbara Seyock, "Trade Ceramics from the Goto Islands (Japan), Circa Sixteenth to Early Seventeenth Century: The Yamami Underwater Site (Ojika) and Related Issues," *Asian Perspectives* 46, no. 2 (Fall 2007): 335.

14. Tōkyō Kokuritsu Hakubutsukan [Tokyo National Museum], *Kareinaru Imari, miyabi no Kyōyaki* [Splendid and refined Imari ware and Kyoto ware ceramics] (Tokyo: Yomiuri Shinbunsha, 2005), 8.

15. Tatsuo Sasaki, "Ceramic Trade via the Oceans," *The Archaeological Journal of Kanazawa University* 67 (2010): 28.

16. Ibid.

17. Ibid.

18. Francois-Bernard Hyghe, "The Kingdom of Silla and the Treasures of Nara (Ancient Korea Kingdom; Buddhist Temple in Nara, Japan)," *UNESCO Courier* (July 1991): 44-46.

19. Ibid.

20. Takenori Nogami, "On Hizen Porcelain and the Manila-Acapulco Galleon Trade," *Bulletin of the Indo-Pacific Prehistory Association* 26 (2006): 124.

21. Nancy N. Schiffer, *Japanese Export Ceramics 1860-1920* (Pennsylvania: Schiffer Publishing Ltd., 2000), 8.

22. Nagatake, *Classic Japanese Porcelain*, 50.

23. Asako Morimoto, "A New Approach to Enigmas from Medieval Hakata: Trade Ceramics as Seen from Archaeological Data," *Guoli Taiwan daxue meishushi yanjiu jikan* [Taida Journal of Art History] 7 (1999): 89–90.

24. Hun-Chang Lee, "When and How did Japan Catch Up with Korea?: A Comparative Study of the Pre-industrial Economies of Korea and Japan" (Center for Economic Institutions Working Paper Series, No. 2006–15, Institute of Economic Research, Hitotsubashi University, February 2002), 10, 22.

25. Schiffer, *Japanese Export Ceramics*, 8.

26. Helen Espir, *European Decoration on Oriental Porcelain, 1700-1830* (London: Jorge Welsh books, 2005), 21.

27. Historically this was spelled: Vereenigde Oostindische Compagnie; in modern spelling this is Verenigde Oost-Indische Compagnie.

28. Paul Doolan, "The Dutch in Japan," *History Today* (April 2000): 36–42.

29. Art Gallery of NSW, "Hasami Ware (Japan): Dish with Incised Waves and Lotus Design," May 2000, http://www.artgallery.nsw.gov.au/work/131.2000/.

30. Nogami, "On Hizen Porcelain," 124.

31. Nagatake, *Classic Japanese Porcelain*, 61.

32. Department for Culture, Media and Sport (DCMS), "Last Chance to Keep 'World Class' Japanese Porcelain from Princess Diana's Childhood Home," February 2011, http://www.culture.gov.uk/news/news_stories/7779.aspx.

33. Christine Moll-Murata, "Guilds and Apprenticeship in China and Europe: The Ceramics Industries of Jingdezhen and Delft" (Paper presented to the S.R. Epstein Memorial Conference: "Technology and Human Capital Formation in the East and West," June 18–21, 2008), 6, http://www2.lse.ac.uk/economicHistory/Conferences/Epstein%20Memorial%20Conference/PAPER-MollMurata.pdf.

34. Ibid., 6–7.

35. Nagatake, *Classic Japanese Porcelain*, 53.

36. Schiffer, *Japanese Export Ceramics*, 8.

37. Nagatake, *Classic Japanese Porcelain*, 54.

38. Nogami, "On Hizen Porcelain," 124.
39. John S. Guy, *Oriental Trade Ceramics in South-East Asia, Ninth to Sixteenth Centuries* (Oxford and NY: Oxford University Press, 1990), 56.
40. Nogami, "On Hizen Porcelain," 124.
41. Ibid., 125.
42. Thi Ha Thanh Nguyen, "European Trade on the Far East and the Mercantile Relationship with Vietnam from the 16th to 19th Century," *Higashi Ajia bunka kōshō kenkyū* [Journal of East Asian Cultural Interaction Studies] 2 (March 2009): 356, http://www.icis.kansai-u.ac.jp/data/journal02-v1/24_nguen.pdf.
43. Ibid., 359.
44. Schiffer, *Japanese Export Ceramics*, 10.
45. Moll-Murata, "Guilds and Apprenticeship in China and Europe," 6.
46. Ibid.
47. Hoshimi Uchida, *Short History of the Japanese Technology* (Mitaka, Tokyo: The History of Technology Library, 1995), 28, http://www.ied.co.jp/isan/sangyo-isan/uchida.PDF.
48. Moll-Murata, "Guilds and Apprenticeship in China and Europe," 8.
49. Shi Yongan, *Riben gutaoci* (Changchun: Jilin Meishu Chubanshe, 1992), 61.
50. Historically this was spelled: Vereenigde Oostindische Compagnie; in modern spelling this is Verenigde Oost-Indische Compagnie.
51. Sasaki, "Ceramic Trade via the Oceans," 29.
52. Oliver Impey, *The Early Porcelain Kilns of Japan: Arita in the First Half of the Seventeenth Century* (Oxford: Clarendon Press, 1996), 81.
53. Miki Sakuraba, "Evolution of Porcelain Tea Ware in Europe From Hizen Porcelain Tea Ware to European Porcelain Tea Service," trans. Jenine Heaton, in *Cultural Reproduction on its Interface: From the Perspectives of Text, Diplomacy, Otherness, and Tea in East Asia*, ed. Nishimura Masanari et al., The International Academic Forum for the Next Generation Series 1 (Osaka: Institute for Cultural Interaction Studies, Kansai University, 2010), 228, http://kuir.jm.kansai-u.ac.jp/dspace/bitstream/10112/3387/1/4-2_SAKURABA.pdf.
54. Ibid.
55. Seattle Art Museum, *Guide to the Porcelain Room: Seattle Art Museum* (Oregon: CDS Publications, 2007), unpaginated.
56. John Ayers, Oliver Impey and J. V. G. Mallet (with contributions by Anthony du Boulay and Lawrence Smith), *Porcelain for Palaces: The Fashion for Japan in Europe, 1650–1750* (London: Oriental Ceramic Society, 1990), 20.
57. Chu Lung-hsing, "The Meeting of China, Japan and Holland: Dutchmen in Japanese Prints during the Edo Period" (paper presented at the Design History Society Annual Conference, University College Falmouth, UK, September 3-6, 2008), 18–19, http://rd.cycu.edu.tw/RD2008/H2500/rtd/project6/97%5C26%E6%9C%B1%E9%BE%8D%E8%88%88ok%5CPaper003_The%20meeting%20of%20China%20Japan%20and%20Holland.pdf.
58. Ibid., 19.
59. Ibid.
60. Ibid.
61. Gisela Jahn, *Meiji Ceramics: The Art of Japanese Export Porcelain and Satsuma Ware, 1868-1912*, trans. from the German, ed. Michael Foster (Stuttgart: Arnoldsche, 2004), 49.
62. Sakai Takashi, "Indoyō no tōji bōeki: Toruko to Higashi Ajia no kōryū o megutte" [The ceramic trade of the Indian Ocean: Concerning exchange between Turkey with the Eastern Part of Asia], *Jōchi Ajia gaku* [The Journal of Sophia Asian Studies] 23 (2005): 261.
63. John Stevenson and John Guy, *Vietnamese Ceramics: A Separate Tradition* (Chicago, IL: Art Media Resources with Avery Press, 1997), 51.
64. Evelyn S. Rawski and Thomas G. Rawski, "China's Economic Development and Global Interactions in the Long Run" (paper prepared for the Harvard-Hitotsubashi-Warwick Conference "Economic Change Around The Indian Ocean In The Very Long Run," Venice, July 22-24, 2008, 4, http://www2.warwick.ac.uk/fac/soc/economics/news/forums/conferences/econchange/programme/rawski_-_venice.pdf.
65. Ibid., 4–5.

66. Renchuan Lin, "Fukien's Private Sea Trade in the 16th and 17th Centuries," in *Development and Decline of Fukien Province in the 17th and 18th Centuries*, ed. Eduard B. Vermeer (Leiden; New York: Brill, 1990), 192.

67. Lin Wei-sheng, "The VOC Trade in Taiwan," *IIAS Newsletter: Research and Report* 31 (July 2003): 30.

68. Lin, "Fukien's Private Sea Trade," 200.

69. Ubaldo Iaccarino, "Manila as an International Entrepot: Chinese and Japanese Trade with the Spanish Philippines at the Close of the 16th Century," *Bulletin of Portuguese/Japanese Studies* 16 (2008): 78.

70. Stevenson and Guy, *Vietnamese Ceramics*, 52.

71. Nan Kyi Kyi Khai, "Myanmā bōeki tōjiki no kenkyū: Marutaban tsubo to hakuyūzara o chūshin ni" [The study on Myanmar trade ware: Martaban Jar and white dish], *Kanazawa daigaku kōkogaku kiyō* 30 (March 2009): 101.

72. Ibid., 101–2.

73. Ibid., 102.

74. Stevenson and Guy, *Vietnamese Ceramics*, 52.

75. Nan, "Myanmā bōeki tōjiki no kenkyū," 105.

76. Sasaki, "Ceramic Trade via the Oceans," 29.

77. For further reference, please see: Arita VOC, Hāgu City Art Museum and Nishi Nippon Shinbunsha, *Tōji no tōzai kōryū ten: Arita, Derufuto, Chūgoku no sōgo eikyō* [East-west exchange of ceramics: Mutual influence between Arita, Delft and China] (Saga-ken, Nishimatsuura-gun, Arita-machi: Arita Vī Ō Shī, 1993).

78. Moll-Murata, "Guilds and Apprenticeship in China and Europe," 7.

79. Ibid., 18.

80. Jahn, *Meiji Ceramics*, 69.

81. Jan de Vries, "The Limits of Globalization in the Early Modern World," *The Semi-Daily Journal of Economist J. Bradford DeLong* (October 17, 2007).

82. Geoffrey A. Godden, *Oriental Export Market Porcelain and Its Influence on European Wares* (London: Granada, 1979), 15.

83. Jahn, *Meiji Ceramics*, 53.

84. Nagatake, *Classic Japanese Porcelain*, 58.

85. Nan, "Myanmā bōeki tōjiki no kenkyū," 31.

86. Ibid., 31, 42.

87. Ibid., 46.

88. Kana Imakiire, "Japonisme in the 19th Century and Western Tableware Made in Japan," *Daisankai Kokusai Nihongaku Consortium* (Tokyo: Ochanomizu University, 2008), 195.

89. Yūko Imai, "Changes in French Tastes for Japanese Ceramics," *Japan Review* 16 (2004): 102.

90. Schiffer, *Japanese Export Ceramics*, 11.

91. Jahn, *Meiji Ceramics*, 18.

92. Imakiire, "Japonisme in the 19th Century," 195.

93. Jahn, *Meiji Ceramics*, 23.

94. Lisa Rotondo-McCord et al., *Imari: Japanese Porcelain for European Palaces: From the Freda and Ralph Lupin Collection* (New Orleans New Orleans Museum of Art, 1999), 9.

95. City Art Museum of Saint Louis, *200 years of Japanese Porcelain*, introduction and catalogue by Richard S. Cleveland (Saint Louis: City Art Museum of Saint Louis, 1970), 9.

96. Koyama Fujio, *Nihon tōji no dentō* [Traditions of Japanese ceramics] (Kyōtō: Tankō shinsha, 1967), 185.

97. Eiichi Yamashita, "Ceramics: The Seto Ceramics School," in *Vocational Education in the Industrialization of Japan*, ed. Toshio Toyoda (Tokyo: United Nations University Press, 1987), 100.

98. Yao Chun, "Tanjiu Riben taoci yaolu jishu" [Investigating Japan's Kiln Techniques], *Zhongguo taoci* [Chinese Ceramics] 143, February 23, 2011, http://enewspaper.ju51.com/xbcy/XBTC/data/20110223/html/30/contentpage_1_0.html.

99. According to Yao Chun's interpretations, such coal-fired kilns were adapted from Northern China' Mantouyao (Bun-shaped kilns). See Yao, "Tanjiu Riben taoci yaolu jishu."

100. Yamashita, "Ceramics," 104.
101. Ibid.
102. Foundation of the Shigaraki Ceramic Cultural Park, ed., *Meiji no yakimono: bunmei kaika no yakimono, Ō-bei o fūbishita japonizumu e* [Japonisme for the Western world: Pottery of the Meiji era] (Kyōtō: The Foundation of the Shigaraki Ceramic Cultural Park, 1996), 3.
103. Yao, "Tanjiu Riben taoci yaolu jishu."
104. This point was argued in Justin Dargin and Tai Wei Lim, *Energy, Trade and Finance in Asia: A Political and Economic Analysis* (London: Pickering & Chatto, 2012).
105. This point was argued in Justin Dargin and Tai Wei Lim, *Energy, Trade and Finance in Asia: A Political and Economic Analysis* (London: Pickering & Chatto, 2012).
106. Yamashita, "Ceramics," 93.
107. Itani Yoshie, "Meijiki ni okeru Nihonjin no shokuseikatsu no hensen to yushutsujiki nitsuite" [Westernization of Japanese food culture and export porcelain in the Meiji era], *Yūgō bunka kenkyū* [Cultural Fusion Studies] 6 (October 2005): 73, 76.
108. Jahn, *Meiji Ceramics*, 31.
109. Schiffer, *Japanese Export Ceramics*, 11.
110. Itani, "Meijiki ni okeru Nihonjin," 76.
111. Jahn, *Meiji Ceramics*, 25.

Chapter Two

Interpretations of Japanese Modernity

A Case Study of Japan's Energy Transition in the Ceramics Industry

Continuing the narrative from the previous chapter on pre-modernity, this chapter will examine the Japanese ceramics industry's historical experience in the modernization process. The chapter uses a variety of sources to study the multiple sources of historical discourses and interpretations on evolving energy use in Japan, particularly in its representative Arita ceramics industry. Academic sources make up an important set of literature on scholarly discourses on the subject matter. The interpretive space is multidisciplinary in nature with the inclusion of perspectives from world history, area studies, Japanese studies, management perspectives, archeology and history. Primary textual materials for interpretation include online websites of local city authorities such as those of Seto City in Japan, trade promotional sites and historical publications (some of them out-of-print) on the subject matter.

The ceramics-making industry is probably one of the oldest manufacturing industries of Japan that survived from the pre-modern age to modernity. Starting with the importation of continental Northeast Asian kiln technologies (in particular China and Korea), the ceramics industry became more sophisticated in making products from earthenware to stonewares and then eventually, fine porcelain wares. The focus of this chapter is on the last item, mainly porcelain products manufactured by the Arita region. Due to increased demand from large external markets like Europe in the seventeenth century onwards, Arita increased its porcelain exports to fill up the vacuum left open by China's Jingdezhen, which experienced internal turmoil in the late seventeenth century.

While late pre-modernity oversaw the growth and expansion of the export porcelain industry, after Japan embarked on her Westernization and modernization programs, the ceramics industry produced porcelain dinnerware made with greater technical precision, consistency and standardization. The production of Western dinnerware porcelain sets were facilitated by the influx of mass production techniques and technologies into Japan from the West. The porcelain-making sector increased its productivity and efficiency made possible by the use of more efficient fuels. As the original location of fine porcelain wares made in Japan, Arita is an important empirical case study in this chapter with some comparative references to other latecomer regions like Seto that caught up with technical progress, public awareness and scale of production.

CHANGES IN THE USE OF ENERGY FUELS

In this section on energy transition in Arita, two main arguments are made. The Arita porcelain industry is part of an industrial trend which sees major industries transitioning from wood-fired technologies to fossil fuels for mass production. The second argument is that, with the transition to higher energy-value fuels, more productive output is possible and at the greater precision level particularly needed for manufacturing Western-style dinnerware sets. The methodology utilized for this chapter is based on a trilateral approach. First, interpretive history is used. Textual information is divided into academic materials (scholarly journals), practitioners' texts (ceramic-making manuals, societies' reports) and a field work section on cave kiln (*anagama*) observation.

According to Christine Moll-Murata, the evolution of energy use in the ceramics industry of Japan evolved from wood-firing a limited volume of ceramics to the systemic use of fossil fuels like coal, oil and gas (imported into Japan in the decade of the 1950s) for mass production.[1] The gradual evolving use of these fossil fuels reflected the same trends found in other industries that require energy to power their activities.

Three qualities appear to stand out in the evolutionary use of fuel that are also generally applicable to the development of other industries as well. First, there is an increase in caloric value of fuels when comparing wood to fossil fuels. In the ceramics production industry, this translated to the capability of firing larger quantities of porcelain wares for exportation. Second, the environmental implications of using these fuels is related to the debate between folk potters and mass production advocates over the use of "natural" resources like timber and wood versus industrial technologies like gas and coal kilns with usage of chemical glazes. Third, mass production required precise work, standardization of procedure and consistency in guaranteeing quality and fine constitution of products.

With regards to the first point, energy use began with the brute force of human labor to pound the clay and the use of timber in the combustion chamber of the kilns found in the ceramics-making industry. Like other industries, coal and oil arrived in the modern period for the ceramics industry. During the Industrial Revolution, coal energy was utilized for firing temperatures high enough for the steel industry and, later, oil was utilized for the internal combustion engine which was one of the backbone systems of the Industrial Revolution. But both fossil fuels produced carbon emissions which were polluting the environment and so natural gas was seen as a cleaner fossil fuel alternative and came into the picture, particularly in the 1960s and 1970s.

In the Japanese ceramics-making industry specifically, gas as an energy resource came relatively late in the 1970s. This decade coincided with the emergence of the environmental and energy conservation movements in Japan which were sparked off by some high-profile environmental incidents and the 1973 oil crisis. While modernization theories focused on utilizing the caloric value of oil and coal for more effective mass production, post-modernist theories looked at the environmental sustainability of production processes and its ability to support middle-class lifestyles minimizing health and environmental problems. This latter priority was important for sustainable development in contemporary East Asia because it met the aspirations of people (particularly the increasingly vocal middle classes) yearning for cleaner environments while leading healthy lifestyles and enjoying the amenities/facilities/infrastructure that contemporary civilization offers and powered by "clean" and green sources of energy. Seen in this global context, Japanese consumers are not unique in their expectations. In societies found within Japan's neighboring region like China, environmental integrity and sustainable development are linked to social harmony and cohesion.

In the case of the ceramics-producing industry, ironically, post-industrial and post-modernist ideas gave rise to a folk pottery movement (*mingei*) that was more interested in the "natural-looking" pottery made by utilizing "natural" non-chemical glazes, timber, and simple craftsmanship to design and create pottery pieces. This movement was a form of counter-culture to the mass production techniques which had been driven by coal and oil-powered kilns for most of late pre-modern and modern Japanese development. It was mostly interested in aesthetic values rather than in productivity, precision and standardization thereby rejecting Tayloristic and Fordist attitudes which fuelled Japan's early modernization process.

There are, however, limitations to this movement because it was unable to completely replace industrial demand for producing ceramic parts and components for the electronics, construction and other industries. The current situation is marked by the co-existence of modern industries with post-industrial/post-modern ideas about environmental sustainability and energy efficiency. The latter is manifested by a return to the use of natural ingredients in production practiced by some potters that I met in my fieldwork and having environmental concerns embedded in ideas about ethical consumption and production. Another influential point is that market demand acts as a kind of economic pressure to bring about high quality standards.

At the beginning of the energy evolution, wood and timber were the earliest fuels used for pottery-making but it involved hard labor. According to Brian Moeran, families engaged in such work needed a minimum of seven days accumulating timber pieces from high-lying areas before chopping and desiccating them for use in the kiln as fuel.[2] Historically, labor-intensive manual firing of pottery can already be observed in the Neolithic period; here one may think of the late Jōmon evidence from Kasori B, dug up in the Kantō region (within the Tokyo Bay vicinity). The simple production techniques of these early times did not automatically imply that the final product was of poor quality.[3] In the context of early primitive cultures and their technologies, their products displayed acceptable quality standards relative to other pottery products found in their age and time. Technologically, however, these were lower-temperature fired pottery like earthenware and clay pottery and not the higher-temperature fired porcelains. Earthenwares tended to be more porous, susceptible to absorbing moisture from the food they hold and crack easily, unlike fine bone china porcelain. In other words, the benchmark for quality standards evolved with changing times, with dynamic market-driven needs and consumer expectations for products that are comparable with their peer competitors available in the markets.

Kilns that fired stone wares rather than fine ceramics were the transitional midway point in the development of kiln technologies between those that made primitive earthenware and the modern sophisticated porcelain-making industry. Stone wares are fired at comparatively higher temperatures of 1,100 to 1,400 degrees Celsius,[4] compared with the more primitive earthen wares. The equipment and knowledge needed to fire stone wares originally came from continental Northeast Asia before the first AD 1,000 years and then spread to the rest of Asia, supported by a network of small-scale ceramics producers which probably constituted the largest proto-industrial clusters in pre-modern world history.[5] Smaller or medium-sized production facilities could support sizable amounts of stoneware output, given the availability of skilled manpower in Northeast Asia at a time when competitiveness was still based on human labor input rather than technological factors related to productivity.

Fired at higher temperatures, porcelain is defined as a bone-white-colored, tough but brittle material made from a combination of substances like kaolinite/feldspar/silicon and baked at more than 1,300 degrees Celsius.[6]

Fine porcelain represents the epitome of firing techniques (not aesthetics), given its thin and delicate appearance and comparatively more uniform composition. In terms of porcelain designs, there are culturally-derived preferences of Japanese ceramics aesthetics by different groups of consumers. For example, even though there are a variety of methods and processes for manufacturing porcelain, the French historically preferred to use white bone porcelain with enamel drawings.[7]

There were also subjective Japanese perspectives when it comes to defining the aesthetic styles that Arita represents. When the author gave a presentation on this chapter topic, one of the audience members insisted that it was not possible to refer to fine porcelain without detailing the Nabeshima variety. Nabeshima was named after a family that built an Arita kiln which fired fine precision porcelain such as table dinnerware and display items that were given as presents at prestigious events to important guests.[8] The producers of Nabeshima kept much of the technology to themselves to protect their share of the lucrative porcelain trade against other families but eventually their design styles inspired potters based in other parts of Japan and the technologies they owned proliferated to other locations in Japan such as Kutani (Ishikawa), Himetani (Hiroshima) and later to Ehime, Fukuoka, Miyazaki and Shimane in the second half of the eighteenth century.[9]

Some of the older Arita celadon pieces were in fact similar in makeup to their contemporary counterparts mass-produced in the Arita gas kiln, fired at 1,300 degrees Celsius.[10] This showed that, while it is possible to fire porcelains with manual labor after such techniques were introduced from continental Northeast Asia, productivity was lower than contemporary kilns, so mass production to the scale that modern kilns were engaged in was not possible and quality of the products was not consistent.

The contestation between traditionally-fired folk wares versus mass-produced high quality wares lasted many years beyond the postwar period. In early postwar days, manual labor in smaller workshops continued to be the way Arita and other pottery-making towns organized themselves for making export ceramics. In the 1950s, according to Steven Tolliday and Yonemitsu Yasushi, Japan was dependent on its exports, particularly to the United States market, because it needed foreign currency to promote economic growth and post-war reconstruction. Arita relied on manual labor for some production-related processes, for example, *kamataki* workers operated and inserted timber and fuel into the timber-fueled kilns. Many of these workers had red-noses and tanned faces from working near high-temperature fires.[11] With the advent of newer kiln technologies that did not use timber, traditional ways of producing ceramics are mostly relegated to heritage or folk pottery movements today.

Newer and more cutting-edge kiln technologies would not be possible without the use of more efficient fuels for mass production but they came with their own set of critics in the realm of the aesthetics. Some potters believe that using polluting fuels (fuels with high carbon emissions) can disturb the potter's harmonious balance with the natural environment.[12] However, for the industries, the pressures for mass production outweigh the concerns of potters who advocate a return to traditional folk techniques in making potteries and ceramics. Despite nostalgic yearning amongst some potters for firing ceramics in the traditional wood-fired way, the progressive use of newer types of fuels continue to be applied in advanced manufacturing processes. Timber-fired kilns were relegated to nostalgic and heritage ventures, particularly amongst folk potters (*mingeika*) making limited quantities of highly-creative and individualistic folk art potteries. Consumers who use their products engage in nostalgic reminiscences of the simplicity of pre-industrial societies.

Other groups of advocates keen on the return to timber energy focused on aesthetic reasons such as the appearance of wood ash on ceramics pieces.[13] Eva Kaminski's paper highlights the aesthetics of Japanese wood-fueled firing process where the singular space inside the kiln permits the fire to engulf the interior walls of the kiln, resisted only by the presence of the porcelain item, resulting in the accumulation of ash on the body of the pottery and the decorative veneer emerging during the firing process.[14] Nostalgia continued to motivate some folk potters to rely on timber-fueled kilns, for example, Takatori Seizan (1908–1982), a descendent of the Takatori pottery-makers with a long tradition of pottery-making since the Tokugawa times, was a leading proponent of reconstructing such kilns to preserve their ancestral traditions.[15]

The smaller pottery-making community of Sarayama, still fired ceramics in the traditional way and they faced the competition from the products made by the community in the nearby Hita region as well as mass-manufactured Arita porcelain that was superior in quality, weighed less, was thinner in constitution, and affordable.[16] Potteries made by modern kilns and firing techniques seemed better than what traditional kilns in Sarayama's backyard kiln could produce. While it took time and catered to particular niche crowds of customers, the desire for hand-made, environmentally-friendly and individually-designed folk handicrafts with Japanese designs found overseas enthusiasts.

On July 2, 2011, on a field trip to Kagoshima Japan, the author visited a wood-fired cave kiln in Mr. Miyagami's hydrangea (*Ajisai*) flower plantation located in the mountainous sections of Izumi city. My host and I had to hike quite a bit to reach the plantation on the top of the mountain. The environment is beautiful with pine trees. The owner believes in the importance of nature and, in his plantation, there is a pottery exhibition area where he displays self-made and hand-designed pottery using clay from Tanegashima. Furthermore, to be environmentally friendly, the owner employs timber from the area and not oil which he believes will emit more carbon. He wants to revive the use of cave kiln which is operated one thousand years ago, to make pottery adjusted to contemporary designs.

While not espousing ideas of the folk pottery movement, Mr. Miyagami's aspirations matched those of the movement with its accent on returning to nature (using timber as fuel, for example), rusticity, simplicity and elegance. He considered these characteristics as elementary standards for technology and design, and on this basis he thought of making non-standardized and non-uniform products of high quality and unique aesthetic values. In his show of reverence to nature, there is an altar at the kiln to offer sake to the kiln god. The host Mr. Tanaka and hydrangea plantation owner Mr. Miyagami both confirmed that this practice is related more to local traditions rather than a strictly Shinto practice. The rites are performed to ensure the smooth firing of pottery.

Even if traditional methods can produce high quality utilitarian and functional kitchen wares, the level of precision is still not adequate to meet industrial standards. According to Charly Iten, one common defect is cracks in the bodies of the porcelain, which are usually due to the firing process known as "kiln wounds" (*kamakizu*) and can be divided into "crack" (*hibiware*) and "hairline fracture" (*nyū*).[17] Such defects needed to be minimized in industrial production. Precise firing for some industrial applications requires the minimization of minor faults that may not be easily achievable with manual firing processes. Another common challenge is distortion. For example, according to the important research on distortion prediction of ceramics by Soejima Kiyoshi and Tomimatsu Kiyoshi, ceramics get out of shape in the porcelain-making processes of shaping, drying, shrinking, and firing.[18] Distortion by itself can also be a form of aesthetics but would not meet the requirements of industrial-standards products like roof tiles and insulators or standardized thin-bodied dinnerware sets.

Besides new technical specifications, higher levels of productivity were needed to manufacture more value-added industrial items such as insulators that can also resist heat[19] and keep the kiln warm with less use of energy. The industry moved beyond only making items of aesthetic value and included those wares with functional utilitarian value as well. These industrial applications require greater precision and harmonization of standards. Some historical statistical figures related to the volume of industrial products made in Arita are listed in table 1.1 in chapter one to give an idea of the increasing scale of industrial application of porcelain. Within the Kyushu region, other pottery-making towns besides Arita began to install better kiln technologies and other techniques to produce porcelain wares. According to the website of the City of Seto, at the turn of the 1800s, Seto was lagging behind Arita technologically in making the surface of the porcelain white in appearance but they gradually received better kiln technologies such as down-draught (*marugama*) and ash glaze (*isubai*)[20] from other parts of Kyūshu. The relevant kilns were installed on slopes with compartments ascending over one another[21] and they consumed traditional fuels like wood and timber.

The revolution in fuel use was a milestone in Seto's ceramics-producing history. According to Steven Tolliday and Yonemitsu Yasushi, traditional timber-fired kilns suffered from disadvantages such as: (1) less accurate firing relative to coal-fired kilns efficiency; (2) 50-60 percent success to make dinner sets with fine quality and bone white color; (3) an average utilization rate of merely two to three firings monthly, while coal/oil-fired kilns are used four to six times and gas kilns ten to twelve times during one month or even on a daily basis.[22]

German chemist Gottfried Wagener spent nearly two and a half decades in Japan and revolutionalized fuel use in the ceramics industry by building the first modern coal-powered kiln.[23] From 1897 to 1906, with the modernization of Japan's energy usage in kiln-firing and the transition from timber-fired to coal-fired technologies with state assistance, the Seto Ceramics School was established in 1902.[24] This made it possible for Japanese entrepreneurs and industrialists to establish larger manufacturing facilities. It encouraged domestic cottage industries to enter the age of mass production by reducing production costs through increased output. Similar observations apply to tiles and dinner-sets that were manufactured in large quantities.[25]

Japan embarked on a program of Westernization and modernization for creating a "prosperous nation, strong army" (*fukoku kyōhei*) in order to fend off colonization. The ceramics-making industry was no exception. The Japanese state deliberately displayed products made by their efforts in modernization at the Philadelphia and Paris Expositions (1876 and 1878 respectively). The new kiln equipment installed in Japan was powered by coal, which, according to Ueki Mikio, further enhanced the already-high pre-modern Japanese standards of ceramics-making.[26] These innovations and the use of modern fuels were designed to signal to the rest of the world that Japan intended to match the West in civilizational attainments and to display its early modernization efforts to stronger nations.

One reason why coal was used in the porcelain-making region of Arita during the early modern period, was that this raw material was available in several nearby coal-mines.[27] According to Yamashita Eiichi, coal-powered kilns made it possible for the porcelain industry to produce new varieties of folk pottery. Some Arita industries turned to making precise and accurately-fired bone porcelain dinner sets used in North America as well as industrial products like tiles.[28] While coal energy might have been a major source of power in the early postwar years, like other industries, the ceramics industry made a transition to oil and gas. In the 1970s, the Germans became the first pottery-makers to use natural gas-powered kilns and this was followed subsequently by the Japanese.[29] Compared to coal, gas was friendlier in terms of environmental impact with lower emissions although natural gas facilities probably required a more expensive capital outlay and the installation of new equipment.[30]

In the environmentally-conscious decade of the 1970s, natural gas was a welcomed alternative to coal. Coal, probably the least technologically-advanced fossil fuel of choice in emerging and developing economies, has its own disadvantages compared with natural gas. It released polluting emissions into the environment. Several photos in the history section of the town's homepage are related to energy use in Arita's past.[31] One set of monochromatic images dating back to the early modernization era depicts Arita with a smokestack against the backdrop of hilly grass-slopes covered by steam coming out from the chimney. The effectiveness of gas-powered kilns is detectable in comparison with its coal-powered counterpart. Coal-powered kilns produce ceramics with a 70 percent success rate; for gas-powered kilns this rate goes up to 90 percent.[32] Gas kilns were advantageous because of the following reasons: (1) at the prevailing level of technological development of kilns in the postwar years, coal was a polluting resource, therefore gas and electric kilns were popular from the 1960s onwards because they emitted less pollution; (2) natural gas kilns enjoyed even higher (95–97 percent) success rates in porcelain output; (3) natural gas kilns were more ergonomic in design.[33]

Like coal, heavy oil kilns also gave off carbon emissions that caused pollution. According to Yamashita Eiichi (whose writings on the Seto ceramics industry was featured in a United Nations University series on the Japanese experience in industrialization), Seto ceramic industries' postwar utilization of heavy oil covered the entire town in black dust, soot and smoke.[34] But the main drawback of gas kilns in the early postwar era was their high cost. Gas kilns remained out of reach for many, especially those in developing regions. Japan's capacity-building experience is instructive. When Vietnamese villages in Bat Trang brought in gas-powered kilns from Japan in 1997, they were equivalent to US$20,000–25,000 per piece which was still out of reach of traditional coal kiln users.[35] It will take some time for kiln technologies that run on gas to become accessible to users in the developing economies as a means to environmental conservation.

NOTES

1. Moll-Murata, "Guilds and Apprenticeship in China and Europe," 22.
2. Brian Moeran, *Lost Innocence: Folk Craft Potters of Onta, Japan* (Berkeley: University of California Press, 1984), 127.
3. Mark E. Hall, "Pottery Production during the Late Jomon Period: Insights from the Chemical Analyses of Kasori B Pottery," *Journal of Archaeological Science* 31 (2004): 1439.

4. Peter Grave, "Melting Moments: Modelling Archeological High Temperature," in *New Directions in Archaeological Science*, ed. Andrew Fairbairn, Sue O'Connor and Ben Marwick (Canberra: ANU E Press, 2011), 216.

5. Ibid.

6. John and Mable Ringling Museum of Art, *Crosscurrents of Design: Asian Export Ceramics at the Ringling A Resource Guide* (Florida: The John and Mable Ringling Museum of Art, n.d.), 22.

7. Imai, "Changes in French Tastes for Japanese Ceramics," 104.

8. Elizabeth Benskin, *The Arts of Japan: A Teacher's Guide* (Washington: Smithsonian Freer Gallery of Art and Arthur M. Sackler Gallery, 2005), 47.

9. Hitoshi Kobayashi, "History of Oriental Ceramics: The Evolution of Japanese Ceramics," The Museum of Oriental Ceramics, Osaka, 2009, accessed March 15, 2012, http://www.moco.or.jp/en/intro/history_j.html.

10. Hiroaki Katsuki et al., "Some Properties of the Early Arita Celadon," *Journal of the Ceramic Society of Japan* 119, no. 1392 (August 2011): 673.

11. Steven Tolliday and Yasushi Yonemitsu, "Microfirms and Industrial Districts in Japan: The Dynamics of the Arita Ceramic-ware Industry in the Twentieth Century," *Journal of Japanese Studies* 33, no. 1 (2007): 49–50.

12. Brian Moeran, "Japanese Social Organization and the Mingei Movement," *Pacific Affairs* 54, no. 1 (1981): 51.

13. Patricia Pelehach, *Dancing with Fire: Building an Anagama Kiln and an Artistic Community* (Pennsylvania: Catskill Arts Society, 2006), accessed March 15, 2012, http://mocany.org/learnKiln.html.

14. Eva Kaminski, *The European Encounters with Japanese Art: Germany as a Case Study, International School Intensive Course: Comparative Cultural Exchange* I (Osaka City University, 2008), 9, accessed March 15, 2012, http://educa.lit.osaka-cu.ac.jp/~ggp/nakami/2008/2008ISIC_Kaminski_paper.pdf.

15. Andrew Maske, "A Brief History of Takatori Ware," The University of North Carolina at Chapel Hill, accessed March 15, 2012, http://www.unc.edu/~mpitelka/takatori/index.html.

16. Moeran, *Lost Innocence*, 32.

17. Charly Iten, "Ceramics Mended with Lacquer — Fundamental Aesthetic Principles, Techniques and Artistic Concepts," in *The Aesthetics of Mended Japanese Ceramics*, ed. Herbert F. Johnson Museum of Art and Museum fur Lackkunst (Ithaca: Herbert F Johnson Museum of Art, Cornell University, 2008), 20.

18. Kiyoshi Soejima and Kiyoshi Tomimatsu, "Research on Firing Distortion Prediction and Correction Techniques for Ceramics Design," in *Processing and Properties of Advanced Ceramics and Composites II*, vol. 220, ed. Narottam P. Bansal, Jitendra P. Singh, Jacques Lamon, Sung R. Choi, and Morsi M. Mahmoud (Hoboken, NJ, John Wiley & Sons, 2010), 159–72. doi: 10.1002/9780470930953.ch17.

19. Hiro Izushi, "Conflict between Two Industrial Networks: Technological Adaptation and Inter-firm Relationships in the Ceramics Industry in Seto, Japan," *Regional Studies* 31, no. 2 (18 August 2010): 118.

20. Seto Municipal Center of Multimedia and Traditional Ceramics, "About Seto Sometsuke (Seto)," accessed March 12, 2012, http://www.city.seto.aichi.jp/e/tourist_infomation/municipal/06.html.

21. Seto Municipal Center for Multimedia and Traditional Ceramics, "Exhibition Outline: Old Kiln House (Seto)," accessed March 13, 2012, http://www.city.seto.aichi.jp/e/tourist_infomation/municipal/04.html.

22. Tolliday and Yonemitsu, "Microfirms and Industrial Districts in Japan," 48, 50–51.

23. Clare Pollard, *Master Potter of Meiji Japan: Makuza Kōzan (1842-1916) and his Workshop* (Oxford: Oxford Oriental Monographs, 2001), 54–55.

24. Yamashita, "Ceramics," 104.

25. Ibid.

26. Mikio Ueki, "The Modernization and Japanization of Contemporary Japanese Ceramics," *Journal of Hokkai-Gakuen University* 123 (March 2005): 2, accessed March 12, 2012, http://hokuga.hgu.jp/dspace/bitstream/123456789/1108/1/GAKUEN-123-1.pdf.

27. David Sanger, "Arita Journal; At Hallowed Pottery, the Master's Wheel Is Stilled," *New York Times,* May 1, 1990, accessed March 15, 2012, http://www.nytimes.com/1990/05/01/world/arita-journal-at-hallowed-pottery-the-master-s-wheel-is-stilled.html?pagewanted=all&src=pm.

28. Yamashita, "Ceramics," 104.

29. Yao Chun, "Tanjiu Riben taoci yaolu jishu."

30. This point was argued in Justin Dargin and Tai Wei Lim, *Energy, Trade and Finance in Asia: A Political and Economic Analysis* (London: Pickering & Chatto, 2012).

31. Arita Porcelain Lab, "What is Arita?" accessed March 16, 2012, http://www.aritaware.com/global/.

32. Quach Huy Thong and Ranjith Perera, "Reduction of Environmental Pollution in Handicraft Manufacturing Enterprises by Using Cleaner Technology: A Case Study of the Traditional Ceramic Producing Village in Bat Trang, Hanoi, Vietnam," in *Proceedings: Sharing of Experiences on Air Pollution Abatement in Southeast Asia (Pre-event, Better Air Quality 2006 Workshop)*, Indonesia, Yogyakarta, December 12, 2006 (Pathumthani: Asian Institute of Technology, 2008), 69.

33. Tolliday and Yonemitsu, "Microfirms and Industrial Districts in Japan," 48, 50–51.

34. Yamashita, "Ceramics."

35. Thong and Perera, "Reduction of Environmental Pollution in Handicraft Manufacturing Enterprises by Using Cleaner Technology," 69.

Chapter Three

Energy Transition in Creative Production

Narratives and Fieldwork about Energy Use in the Ceramics-Making Industries of Hong Kong, Japan, and China

Energy transitions for many mass manufacturing industries have transited deterministically from wood and timber to hydrocarbons like coal and then oil and natural gas. Some motivations for making this transition were fuel efficiency, increasing environmental awareness and progress made in ceramics-firing technologies. This chapter is interested to examine narratives related to the deterministic progression of fuel use in the artistic ceramics-making industry. In studying this industry, the chapter is interested in three questions: (1) narratives related to the natural progression of fuel use; (2) the impact of environmental considerations on fuel use; (3) technological trends in the use of fuels.

This chapter concentrates on fuel use for the kiln-firing process rather than pre-firing production stages due to the following reasons. First, the kiln-firing process is a necessary component for all ceramics manufacturing processes regardless of historical time and context. Therefore, it is useful for studying the evolution of fuel use for the industry. Second, transition from the use of timber and wood to fossil fuels is directly related to technological advancement. This is one of the main factors behind changing fuel use. Third, the firing process in the kiln is arguably the most important production stage because it gives ceramics the definitive color and glazes that characterizes the appearances of the ceramics. They are visible results of different techniques used in firing ceramics and decisions made in terms of fuel usage.

The structure of the chapter is as follows. After the introduction, the chapter will discuss the methodology used for this chapter. This is followed by a historical overview and literature review of fuel use in the ceramics industry, the purpose of which is to point out the deterministic march of energy use towards more efficient and environmentally-friendlier fuels. It is then followed by a concluding section discussing some evolving trends in energy use to fire traditional ceramics today.

METHODOLOGY

The chapter discusses how fuel use for kiln-firing is influenced by environmental, productive and aesthetic considerations. It utilizes fieldwork based on conversations, dialogues and interviews with ceramicists, artists and potters.[1] The oral historical method of corroboration is employed here as conversations and oral interviews provided by the artists are integrated with textual sources related to the subject matter. As part of the fieldwork conducted for this chapter, the author has carried out observation work in Hong Kong and Japan, speaking to individual artists and visiting museums as well as galleries. The fieldwork conducted was useful in learning more about electric kiln use in Hong Kong, China as well as Japan. At the museums and galleries, the author collected textual narratives related to the subject matter. Other than fieldwork interviews, Japanese manuals on ceramics-making as well as kiln construction proved to be a rich source of information for the author in the area of fuel use. This may explain why there is an active publishing industry on the topic of kiln construction and firing methodologies for making artwork ceramics. The chapter does not pretend to be comprehensive but presents some historical milestones in the ceramics-making industry in terms of energy use.

Historically, in the context of Northeast Asia, the first modern kilns appeared in Japan near the time of Meiji Restoration 1868. Japan was the first Asian nation in the world to modernize its industrial production system for mass production. Originally, like the Chinese and other Asian ceramics-manufacturing industries, Japanese ceramics manufacturing facilities used traditional wood-fired technologies to produce their porcelain. But German chemist Gottfried Wagner (1831–1892) transformed the industry when he introduced modern coal-fired technologies to the Japanese ceramics industry and became known as the creator of the modern ceramics industry in Japan.[2] According to the Saga Kenritsu Kyūshū Tōji Bunkakan, Wagner constructed the inaugural Westernized coal-fired kiln in Arita in 1870 and supervised the building of another kiln in Tokyo in 1882.[3] Gisela Jahn's important work on

Meiji Ceramics details Wagner's (sometimes spelt as "Wagener") contributions to imparting coal-firing methodologies to the Japanese:

> He supervised construction of a coal-fuelled kiln, the first in Japan and the first in a long sequence of experimental kilns that came to an end in 1902, when a construction with updraught was finally mastered. Not until then did coal-fuelled kilns become widely used in the production of porcelain in Japan, not only in Arita, but also in such other centres as Seto and Tajimi.[4]

Modernity is also signified by the transition from the use of timber-fired dragon kiln to coal-fired kilns.[5] From 1897 to 1906, the Japanese state aided ceramics-making regions with the technological transition to coal-fired kilns after Wagner's pioneering successes, including financing a test-kiln built in the Seto Ceramics School in 1902.[6] Modern coal-fired kilns revolutionized the ceramics industry, allowing smaller-scale folk ware industries to produce highly standardized Western-style dinner table ware as well as other industrial ceramic products such as building tiles.[7] Modern coal-fired technologies also made it possible to have comparable quality control for each individual item. This was important because Western dinnerware sets required accurate replication of standardized patterns and sizes. Meiji Japan's Arita and other porcelain-making centers followed these requirements to manufacture export wares to Europe and obtain foreign credits needed for national modernization. Such manufacturing processes also became important learning experiences for Japan's overall modernization process, especially as it picks up technical skills for moving on to higher value-added manufacturing. Japan's head-start over other Asian industries in this learning experience curve and its culture of continuous innovation allowed her to develop sophisticated electric and gas kiln technologies later in the postwar years, making her a world leader in kiln technologies.

PEOPLE'S REPUBLIC OF CHINA (PRC) AND EARLY POSTWAR STATEHOOD AND MODERNITY

Likewise, coal also became an important energy resource in the PRC's early phase of industrialization. Taking advantage of resource availability, large-scale ceramics production areas were strategically located near the coal production areas, for example, in Tzu-po municipality's Po-shan.

> In [1930s] Po-shan the Japanese owned the local coal mines and had started many small domestic potteries, using coal to fire the kilns. After the war when Po-shan was rebuilt some of the old beehive-shaped kilns were rehabilitated to supply the immediate needs of the local peasants and to start a ceramic industry. In 1955, there were one hundred workers who could set about building up the industry but their first products were only of medium quality. After some

> of the potters had visited the Tang-shan complex in Hopeh Province for further experience, the quality had so improved that by 1959 Tzu-po porcelain was being exported to South-east Asia, Canada, Australia, the USSR, Cuba and Europe. [8]

There are three important points in this quotation. First, there was an existing relationship between Japan's modernity and the PRC's nascent industrialization. Japan's modern coal facilities were taken over by the PRC for their own production. The modern coal mines made it possible to supply early modern PRC kilns with energy for production. Second, quality continues to be a motivating factor for fuel use and technological upgrading in the Chinese context. Like their Meiji Japanese counterparts, the PRC industrial potters needed to standardize the quality of their export wares for meeting the expectations of consumers in developed markets overseas in order to earn foreign credits and currencies to finance their nation-building industrialization process. Third, modern coal facilities left behind by the Japanese and primitive bee hive kilns served as learning experiences for the PRC industrial workers to pick up the necessary technical skills to run larger and more complex manufacturing facilities. Po-shan served as a learning experience for other facilities with the objective of encouraging constant improvements in quality of the product output. All these objectives had strong parallels with the Meiji Japan example. In terms of energy use, Po-shan's case indicated hydrocarbons like coal were preferable as a fuel of choice due to availability of technologies to extract and consume coal resources, at least in this early postwar industrial phase in China.

TRANSITION TO OIL

While coal energy was equally utilized in the pre-modern and modern contexts, the use of oil and petroleum energy was a clear demarcation between pre-modernity and modernity. The progressive march of modernity and modernization did not stop with coal energy use. As some Chinese industries converted to the use of oil as fuels for the mass manufacturing activities, its ceramics industry also made a similar transition. Again, back to the example of Po-shan:

> By encouraging co-operation and exchange of ideas, the great ceramic complex at Ching-te-chen in Kiangsu Province has benefited the now extensive industry at the Po-shan State Pottery. The proximity of Po-shan to the Sheng-li oilfields is reflected in the fact that ten of the fifteen continuous kilns in the potteries are now fired with oil, the rest with coal gas produced in their own factory. Po-shan was the first ceramic complex in China to install continuous kiln. [9]

Just as it took time for diffusion of technologies to take place, due to the large geographical makeup of China, kiln modernization occurred differentially in different parts of the country. In some parts of China, for example, in the Yixing (I-hsing) pottery-making region, conversion from the use of traditional dragon kilns to continuous coal-fired kilns only occurred in the 1960s.[10] As part of the modernization process, energy use in the Chinese ceramics industry experienced the progressive transition from timber and wood to coal and oil. In general, fossil fuel use continued uninterrupted until environmental consciousness and energy shortage led to a rethink on the deterministic progression of energy. Instead of fuel efficiency, cleanliness of fuels became a major concern amongst the industrial energy consumers. Figure 3.1 is the author's diagrammatic representation of the evolution of hydrocarbon use in mass manufacturing industries.

Japan reduced her industries' overreliance on oil use after the 1973 oil crisis when a looming oil embargo threat motivated Japan to introduce energy efficiency into all her industrial sectors. To cope with the oil crisis, Japan retooled her technologies to make them more environmentally-friendly and energy efficient. Consequently, contemporary Japanese kiln-makers developed more energy-efficient and environmentally-friendlier technologies touted as "smokeless" or "environmentally-friendly (*kankyō ni yasashi*)" that run on kerosene but produce the same glaze effects as traditional kilns. The only disadvantage of using kerosene-powered kilns is that they generally give off an odor which may make them unsuitable for use in the household or small studios. Other than decreased levels of waste products and emissions, newer kiln technologies also have the capabilities to minimize noise levels (a disadvantageous feature of kerosene kilns are their noisy exhaust fans). However, the disadvantages of using kerosene fuels did not lead to total disuse of kerosene kilns. Instead of being phased out, continuing interest in oil-fired kiln technologies remains due to the fact that, like coal firing, oil-based kiln technologies allow ceramics to have natural finished glazes.

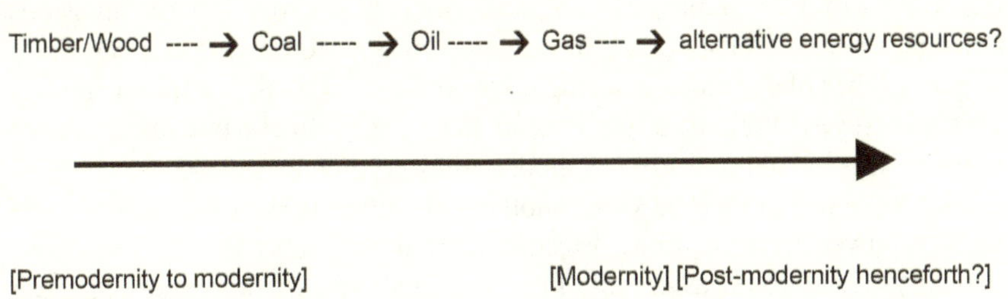

Figure 3.1. Diagram of fuel progression from timber/wood to alternatives (author's own diagram).

Though the Japanese created more energy-efficient oil-fired kilns, some old problems remain because even new-generation oil-fired kilns continue to release carbon dioxide and other emissions into the air. They are also larger and less portable than electric kilns. The environmental aspect of oil and coal use in the ceramics industry became an increasingly important factor of consideration for the artistic community. As the first country to modernize in East Asia, Japan also became the first postwar economy in East Asia to emphasize the environmental aspects of industrialization. Japan experienced the rise of environmental movements led by civil society in the aftermath of a series of high-profile environmental challenges like Minamata in the 1960s and 1970s. Because of increased environmental awareness, Japanese kiln makers considered other alternatives for firing. Environmental reasons led to the development of newer environmentally-friendlier and energy-efficiency kiln technologies such as gas and electric kilns. Newer kilns also have better heat-retentive properties that contribute to efficient use of energy.

TRANSITION TO GAS AND ELECTRIC KILNS

I will first discuss the emergence of gas kiln technologies. According to Yao Chun's evaluation, Japan's lead in porcelain arises from continuous innovation in new types of kiln technologies.[11] Japan adapted German technologies and constructed their own indigenously-designed gas kilns for porcelain manufacturing in the 1970s.[12] In an age of environmental awareness, coal is considered inefficient, polluting and dangerous compared to cleaner fuels like oil and gas, but, between these three fossil fuels, natural gas holds the title of being the cleanest fossil fuel in terms of emissions.[13] The gas revolution took place at a time when emissions from coal and oil-fired kiln plants became a source of concern for environmentally-minded individuals. The pollution created by large-scale coal firing became obvious in some industrial areas for ceramics manufacturing. According to Eiichi Yamashita, the town of Seto used so much coal for firing porcelain that even sparrows flying over the town turned black from the soot pollution.[14] In such cases, natural gas kilns provided a possible solution since it has the lowest emissions amongst the fossil fuels. Gas kilns are also cheaper to operate and can fire a larger number of ceramics in a single session and, like oil, it allows ceramics glazes to reveal their true qualities in the firing process but one problem remains. It is noisier and still has more emissions than electric kilns.

Enter the use of electric kilns, another revolution in the evolution of fuels used to power ceramics kilns. Electric kilns are probably the most environmentally-friendly and affordable kilns for small-scale studio and home use today. The use of electric kilns is also widespread due to the features stated below. Their advantageous features are easy handling, clean odorless emissions, portability, compactness and suitability for urban environments. Electric kilns are also quiet in operation and have easy-to-handle temperature controls. Because of the ease of temperature control, defective pieces are minimized but the chances of obtaining "accidental" glazes that are aesthetically pleasing are minimized as well. Because of the quantity of ceramics fired (and the large size of these kilns) and the difficulties in temperature control, longer duration of firing in traditional timber or coal-fired kilns, the glazes may not conform precisely to one's intended design. The difficulties in operating traditional kilns compared with electric kilns also mean that the latter has a liberating effect for artists with space constraints. The portability, miniature size and ergonomic features of electric kilns facilitate wider participation by interested hobbyists and recreational potters in ceramics-making activities, especially those living in urban areas. Electric kilns are also less costly to operate. In fact, because of the above mentioned reasons, Japan experienced a 100V electric kiln boom in 2005, making it possible for many hobbyists to realize the dream of installing a kiln in the home.[15] My informants, for example, cited the impracticality of setting up large traditional kilns in Hong Kong.

The author's field study of the fourth generation of contemporary Hong Kong ceramicists indicated three advantages in using gas and electric kilns. First, they were cleaner than other fossil fuels. Both types of kilns gave off less waste products and emissions compared with timber or coal kilns. Environmental factors were especially important in Hong Kong and urban environments in Japan because of their densely-populated urban environment. In Hong Kong, for example, the potters' past experiences with burning timber and recycled furniture pieces resulted in sooty smoke and fumes. Hong Kong laws do not allow wood-firing. Second, the idea of a community of people supporting ceramics production activities is a theme that constantly emerges in academic and trade literatures as well as from my oral interviews with contemporary Hong Kong potters. However, as electric kilns were highly portable and fit snugly into urban studios, they were easy to use compared with timber or coal fired kilns which required complicated preparation procedures and took up space for storage. Consequently, when electric kilns are used, there was no need for a community-wide mobilization of human resources for production-related activities. The Hong Kong-based potters also depend on informal networks to obtain their supplies for production. Technological means makes it easy to obtain the required materials for pottery-making through the internet, phone and social media. It does not depend on a closely-knit community of suppliers to serve the needs of potters. In fact, supply lines for such materials are now global and have gone virtual too. Third, they provided ceramicists and artists with greater control and handling to achieve ideal glazes.

The emergence of technologies and tools for making small electric and gas kilns resulted in a boom and proliferation of handmade kilns in Japan. A variety of hand-built kilns can also be found in Japan, including oil-drum kilns. These oil drum kilns are typically lined on the bottom with an insulation blanket and fire-proof bricks that are held together by cement mortar. The drum body is also surrounded with insulation blanket with inlaid felt materials held together by cement mortar. The oil drum cap is similarly lined with the same material. Holes are then made on the cap and the drum is wired up with heating elements and electric wiring. The insulation blanket prevents heat from escaping the kiln body so that efficient firing can be carried out. By comparison, in primitive pottery-making societies, sometimes dried animal dung, pressed mud and pottery covers are used,[16] but heat escapes easily from these materials. Other kinds of small-scale home-fitting compact kilns include the stacked-up coal-fired kilns. They have holes by the side of the kiln like their large wood-fired counterparts that allow air to drift upwards drawn along with the lighter hot air draft. One such model is produced by Kabumatsu Pte Ltd[17] while other makers use an oil drum shaped coal-fired kiln. The secret behind the environmentally-friendlier coal-fired home kiln is to use smokeless or low-smoke coal that combusts efficiently to minimize environmental impacts while firing pieces effectively in a small kiln body. The downside to this approach is that environmentally-friendly high-quality coal may also be comparatively costlier than ordinary coal bits.

Because of the narrow spaces that small compact kilns have to work with, they typically have ergonomic designs that are user-friendly. For example, they have doors that open up like a microwave oven cover or can be lifted up like a cap (some have both features, enabling half of the kiln body to open up from the top and the side). From my fieldwork observation in Hong Kong, most potters prefer a lid that opens from the top for reasons unknown. Regardless of how it opens up, these features make it much easier to retrieve the finished products as opposed to wood-fired kilns. Some of them are even equipped with roller wheels or sliding trays inside the kiln body so that the finished pieces can be slid out. In contrast, the traditional Japanese climbing kiln that are much larger require the operator to wait for temperatures to cool down before entering the kiln to remove the fired pieces. A large wood-fired kiln also requires temperature inside the kiln to cool down before removing the fired clay items from the kiln body through a small opening. In both cases, there are no handles to operate the openings.

To facilitate compactness, there are at least five other design features and functions of contemporary electric kilns that removes the need for a complex team of specialists to operate the technology. First, many contemporary electric kilns now come with numeric soft touch pads that are ergonomically designed for optimal temperature inputs. In other words, they are designed for one-man-operation. The newest models, according to my informants, can even be operated remotely by smart-phones. Second, there are little harmful emissions from electric kiln firing and its gaseous output can be easily dissipated through the use of an exhaust fan. This is a vital point for Hong Kong potters who work and live in narrow spaces while coping with high rentals and property costs. Some of the Hong Kong potters that I interviewed even built their own filtration system for additional precaution before releasing the gaseous output into the atmosphere. Third, they are typically small and portable, and can be transported by a potter (some are built on roller wheels) or with the help of some friends and family members. Fourth, because electricity is piped into the home, there is no need to gather and cut timber. There is also no need to transport these fuels to the firing sites. Lastly, due to the presence of precise controls, chances of accidents occurring in the firing process are minimized, reducing the frequency of operations for the kiln. From a review of kiln manuals gathered during my fieldwork in Japan, I found the price range of contemporary electric kilns (depending on size and functionality) pegged approximately between US$400 to US$55,000. They are typically cheaper to operate than their kerosene and gas counterparts. In Japan, a 100V electric kiln is the typical capacity for a mini-kiln situated in a high-rise apartment. It fits snugly into the living room. This kind of kiln fits into an urban lifestyle and is equally applicable for a densely-populated and crowded city like Hong Kong as well.

Besides design features, other benefits and advantages in using home-fitting kilns include the simple joys of creating something with their own hands. The manual work involved in building the kiln, the experimentation involved and unexpected results of the output are motivations driving these enthusiasts. The hand-made kilns tend to have less precise controls compared with fully-computerized contemporary electric kilns. From my fieldwork and literature review, unexpected glazes and results from firing are part of the joys experienced by potters and ceramicists. One of the exhibition brochures, which a Hong Kong informant gave me, quoted artist S.K. Chan's views on the unexpected outcomes of firing: "Ceramics faults like cracks, explosions, melting, collapsing, over-or-underfiring and glaze running excessively all over the kiln shelves, etc. were of course unavoidable." [18]

These "faults" can be minimized through the skills of professional potters and with the use of electric kilns with their precise temperature control and ergonomic designs for the control panel. While they were listed by the artist in the brochure as "faults," in fact, artists and ceramicists have frequently stated that unexpected results are part of the excitement in firing. The ease of control and portability of small home-built or hand-made kilns enables enthusiasts and hobbyists to pursue this interest with the least amount of preparations and moderate expenditure. For those who prefer off-the-shelf ready-made technologies, other than building a kiln, the other alternative is to purchase a fully computerized, alphanumeric pad-controlled electric kiln mounted on wheels. Like the advent of modern coal-firing technologies which empowered smaller industries to fire export wares and construction ceramic products in the early modern period, the emergence of smaller home-sized kilns and computerized electric kilns also facilitates micro-entrepreneurship if and when hobbyists and amateurs decide to move into the business of making pottery for sale, on a part-time or full-time basis.

CONCLUDING SECTION

While technological, operational and environmental factors motivate contemporary artists to use electric kilns or resort to smaller home-fitting units, large traditional kilns continue to exist and are preserved in Hong Kong, China and Japan. Some artists, government agencies and hobbyists choose to preserve older methods of production or hydrocarbon-operated technologies due to preservation of heritage and traditions and aesthetics. Even though technological progress has made traditional kiln firing methods obsolete, old pre-modern kilns are now preserved as a form of heritage. An example in Hong Kong include the traditional Wun Yiu kiln *Wanyao* (kiln used for making porcelain bowls) in Tai Po (*Dapu*) Road which is preserved as the last remaining traditional kiln in Hong Kong. This heritage site was first revealed to me by my informants in my fieldwork in Hong Kong. There are quite a number of online resources related to this kiln and one of the most informative websites is run by the Heritage Office in Hong Kong. The Wun Yiu embodies many of the points related to traditional climbing kilns. First, it is built on sloping land that connected the villages of Sheung Wun Yiu, Ha Wun Yiu, To Chi Yin, Cheung Uk Tei. The mining facilities located at the facility indicated a specialized division of labor existed in extracting the natural resources needed for production. Wun Yiu is located near "the rich clay mines in the hills, dense vegetation and the abundance of firewood."[19] Its dragon-kiln design needed the use of firewood nearby for firing. Though the technologies at this historical site are made obsolete by modern technologies and contemporary environmentally-friendly kilns, they are consciously preserved by the Hong Kong government as important heritage artifacts.

Wun Yiu stopped operating in 1932.[20] It has become a symbol of heritage preservation. It has become part of the general narrative from Hong Kong's Commissioner for Heritage's Office in linking tourism with heritage. The association of heritage with consumption is a practical strategy to actualize self-sustaining heritage industries. For example, in the heritage newsletter "on revitalization" published by the Commissioner for Heritage's Office, the Wun Yiu was featured in heritage panel displays at the Tai Po Mega Mall and they elicited positive responses from the public. The newsletter contained the following quotation by "Ms. Lau": "I am so surprised we have an archeological site in Tai Po, which is not far from where I live. I definitely will make a visit to Wun Yiu with my family during the weekend." [21]

In terms of policy, there are many different narratives and publicly-articulated ideas when it comes to approaches in Hong Kong's heritage preservation. I will briefly focus on three of them here but a detailed discussion on heritage policies is outside the scope of the chapter. Interested readers can reference detailed works listed in the bibliography. All three narratives are equally valid and have their respective strengths.

In terms of identity, the first narrative promotes a unique Hong Kong identity that juxtaposes Chinese ceramics-making history (neutral of regime association) with development and modernization in its colonial history. This narrative is also market-friendly by supporting the costs of heritage preservation through the generation of tourist revenues. A representative work in this area is Joan Henderson's publication listed in the bibliography.

Another narrative sees the idea of heritage less as a political or economic issue and instead conceptualizes it as a cultural construct. This school of thought is concerned with the imageries of heritage sites which may symbolize different meanings to different groups located at different levels of Hong Kong society (territory-wide, local community or village level). Because heritage is conceptualized as a construct, this school of thought is therefore keen to unravel contesting ideas about heritage for different stakeholders so that there is equitable appropriation of the symbols by all interested parties. One important work in this area is by Sidney Cheung listed in the bibliography. [22]

The third view conceptualizes heritage as a form of cultural tourism. This is an economic perspective of heritage. To maximize the economic value of the cultural and heritage assets of Hong Kong, this school encourages close links and commercial partnerships between stakeholders and parties interested in accessing the cultural asset. The interest here, therefore, is not negotiating the issue of identity between political forces or unraveling constructs to discover the symbolisms behind cultural assets. Instead, the goal is to determine the ownership of those assets, identify them so that management of those assets can be institutionalized (by the state or private sector) and opened to visitors. A detailed account can be found in the publication by Bob McKercher, Pamela S.Y. Ho and Hilary du Cros listed in the bibliography. [23]

These three views are not necessarily mutually exclusive. All three are market-forces driven centered upon the idea of revenue generation. All three narratives also demonstrate pragmatism in navigating different ideas about Hong Kong's identity by emphasizing a consultative approach, particularly amongst parties who have a strong interest in defining cultural assets for their agendas. All three narratives are also cognizant of the contesting forces at work in accessing cultural and heritage assets and therefore do not advocate a uniform approach in managing all sites. These eclectic reasons are all contributive towards the motivation to preserve the Wun Yiu traditional kiln.

Not all conservation efforts originate from the state, the artistic community or other institutionalized platforms. Some of these conservation projects are personal in nature. In the case of Japan, the author visited an *anagama* kiln in Izumi-city in Japan in the summer of 2011 that was reconstructed from ideas related to kiln construction dating back millennia in the Satsuma tradition. It is a personal project funded and built by an individual enthusiast. I have written a separate chapter for this observation study. The construction of a modern *anagama* kiln today typically consists of building a kiln using fire-proof bricks on a natural slope or alternatively building a wooden frame over a piece of flatland before covering up the wooden frame up with bricks. The wooden frames are needed to create the curvature for the kiln cover. When there are no natural slopes available, for example, on flat plains or agricultural lands in Japan, some kiln-builders build soil ramparts to create the sloping angle. The bricks today are generally held together with applied fire-proof mortar. The base of the kiln is sometimes constructed using bricks or concrete. Reconstruction or replication of ancient kilns may be important for heritage because the pre-modern climbing kilns (*noborigama*) in the major porcelain-producing centers of Arita and Okawachi are disappearing physically due to the impact of natural elements. The Kyushu Ceramics Museum noted this phenomenon:

> Climbing kilns (*noborigama*) were rebuilt over and over again, bearing the pounding of many years of flame, wind and rain. However, even vital kilns pouring forth greater amounts of smoke disappear from view when shifts in styles occur, leaving only the traces of their former kiln site.[24]

Satellite pictures and archeological expeditions have facilitated the identification and preservation of these old sites. It is likely to take a combination of state, non-state and local community initiatives to restore and revive these kiln sites and uncover artifacts eroded by natural elements. The statement above indicated one important observation which is highly-visible voluminous emissions from the climbing kilns that are evidential of some form of pollution from using wood and timber fuels in pre-modern production.

Besides understanding about the past, the preservation, reconstruction and replication of older kilns also serve another purpose. As the author reviewed a number of guidebooks on kiln tours or pottery-making hobbyist guides published in Japan, he realized that kiln-tours have become an industry by itself. To attract visitors and tourists, the descriptions of the kilns depicting them as historically-significant sites, e.g., the largest in a region or the oldest kiln in the national, etc. A thriving kiln-tour industry may mean greater prospects of reviving traditional kilns that use wood, timber or other older forms of energy resources. Kiln tour activities may also include retailing souvenirs whose sales contribute to the financial sustainability of local community.

PASSING ON TRADITIONAL SKILLS

Besides saving kiln heritage artifacts, intangible heritage like artisanal skills are also preserved and passed on to succeeding generations of craftsmen. The use of traditional fuels continues to be important for the production of folk pottery (*mingei*) in Japan which uses both coal and red pine (*akamatsu*) fuels. In order to extract these fuels and prepare them for traditional methods of firing, the presence of a community of specialists with specific skills is needed. For example, in Moeran's important study on Sarayama, an entire community is mobilized for preparations needed for the ceramics pre-firing stages, for example, gathering timber, chopping them and stacking firewood for storage.[25] First, wood needs to be cut down by lumberjacks and placed in storage spaces. Then, manpower is required to move the logs from the timber storage area to the kilns. Cutters are then employed to saw these logs into suitable lengths for insertion into the kilns before the potters and kiln operators use the wood as fuel for firing. All these processes are managed within the community and sometimes depend on neighboring regions for supplies. Due to the interdependence of all these procedures, every step has to be done carefully to ensure natural resources are conserved for the continued prosperity of the local economy. Besides communitarian, environmental and aesthetic factors, in the case of Japan's *mingei* folk pottery movement, Moeran[26] also raises the element of nostalgia as a rationale for continuing with older methods of folk pottery production. Moeran's study, however, did not clearly indicate whether the use of natural ingredients represents the community's yearning for nostalgic ways of production as a form of counter-reaction against modernization or if his study simply portrays a peripheral community untouched by urban development and keen to preserve artistic notions of traditionalism.

While community-based production ecologies can be found in rural or suburban potter communities, densely-populated highly-urbanized contemporary environments such as Hong Kong Island's Central district cannot support the same kind of production environment. With the availability of online technologies, materials used in the kilns are now sourced online and delivered via postal mail to the artists. Contemporary Hong Kong potters are able to tap into a global online network and source for materials based on individual needs, rather than relying on a network of suppliers and technicians found within a single local community. Almost all contemporary professional full-time potters in Hong Kong that I met in my fieldwork used this technology to source for their firing ingredients. When electric kiln owners really want to carry out wood and gas firings, they travel to other pottery communities around the world to use their facilities. For example, the fourth generation of Hong Kong contemporary potters revealed to me that their urban electric and gas kilns are unable to achieve certain ash effects found in coal/timber-fired firing. Therefore, they travel to Taiwan or Jingdezhen to use traditional kilns found in those areas to fire their products.

NOTES

1. Interview with the fourth Generation Hong Kong-based potters and ceramicists, Ms. X and Ms. Y. The fieldwork respondents for this chapter are kept anonymous.
2. Yamashita, "Ceramics," 100.
3. Saga Kenritsu Kyūshū Tōji Bunkakan, *Earth and Fire: The Historical Development of Kyushu Ceramics: The Kyushu Ceramic Museum, The Permanent Exhibition Guide Book.* (Saga, Arita: Saga Kenritsu Kyūshū Tōji Bunkakan, 1996), 23.
4. Jahn, *Meiji Ceramics*, 113.
5. Yao, "Tanjiu Riben taoci yaolu jishu."
6. Yamashita, "Ceramics," 104.
7. Ibid.
8. Wanda Garnsey and Rewi Alley, *China, Ancient Kilns and Modern Ceramics: A Guide to the Potteries* (Canberra and Miami, Fla.: Australian National University Press, 1983), 35-36. Please note that "Po-shan" is spelt as "Baoshan" in the Hanyu Pinyin Romanization system.
9. Ibid., 37.
10. Ibid., 86.
11. Yao, "Tanjiu Riben taoci yaolu jishu."
12. Ibid.
13. This point was argued in Dargin and Lim, *Energy, Trade and Finance in Asia*, unpublished at the point of this writing.
14. Yamashita, "Ceramics," 93.
15. Kōichi Noda, *Hajimeteno tōgeigama erabi* [First time to choose kilns] (Tokyo: Seibundō, 2005), 20.
16. Ryū Nagumo, *Tōgei* [Ceramics] (Tokyo: Nichibō Shuppansha, 1998), 9–23.
17. To obtain more details of their coal-fired mini kilns, please refer to: Kabumatsu company's website: http://www.kabumatsu.co.jp.
18. Committee of Dual Plasticity Ceramics Exhibition, *Dual Plasticity Ceramics Exhibition*, Jockey Club Creative Arts Centre, October 22, 2010 – November 6, 2010 (Hong Kong: Committee of Dual Plasticity Ceramics Exhibition, 2010).
19. Commissioner for Heritage's Office, Development Bureau, *A Newsletter on Revitalisation: Revitalisation@Heritage*, no. 25 (Hong Kong: Commissioner for Heritage's Office, Development Bureau, July 2002).
20. Commissioner for Heritage's Office and AMO, "Tai Po Wun Yiu Kiln Site," accessed October 16, 2012, http://www.heritage.gov.hk/taipowunyiu/en/background.html.
21. Commissioner for Heritage's Office, *A Newsletter on Revitalisation*, 3.
22. Cheung, Sidney C.H. "Remembering through Space: The Politics of Heritage in Hong Kong," *International Journal of Heritage Studies* 9, no. 1 (2003): 7–26.
23. Bob McKercher, Pamela S.Y. Ho, and Hilary du Cros, "Relationship between Tourism and Cultural Heritage Management: Evidence from Hong Kong," *Tourism Management* 26, no. 4 (2004): 539–48.
24. Saga Kenritsu Kyūshū Tōji Bunkakan, *Earth and Fire*, 10.
25. Moeran, *Folk Art Potters of Japan*, 99–122.
26. Ibid., 15.

Chapter Four

Blue and White Fired Clay in Everyday Lives

An Ecology of Creative Production, Energy Use, and Aesthetics of Arita (with Comparative References to Jingdezhen)[1]

BACKGROUND

This chapter is made possible with the generous support of the Hop Wai Short-term Research Grant. It is a continuation of the narrative in Chapter One, titled *Japanese Trade Ceramics in the Premodern Global Trading Space* which left off in the pre-modern era before a general discussion on the evolution of energy use in the ceramics industry. This chapter focuses back on the case study of Arita and looks at the micro-study of a popular and important class of porcelain known as blue and whites or *sometsuke* in Japan and *qinghuaci* in Chinese. In Chapter One, I discussed three main points: (1) Japan's Arita porcelain kilns were able to substantially increase their export porcelain to Europe due to dynastic transitions (Ming to Qing dynasties) in China affecting Jingdezhen production; (2) constant innovations allowed Japan to adapt Chinese ceramics manufacturing techniques and indigenize them for designing and creating Arita porcelains along with other continental influence from Korea; (3) Japan embarked on modernization and westernization first in Asia and quickly overtook China in production quality, techniques and mass production numbers. In Chapter One, I focused on a historical survey of mainly blue and white[2] export ceramics from Arita and Jingdezhen from the seventeenth century onwards. I also surveyed the export history of these two types of blue and white ceramics to third parties in East Asia and beyond to consumers markets in Europe starting from the seventeenth century onwards.

In the same chapter, I argued that modernization differentiated the developmental trajectories between the two ceramics production centers in Jingdezhen and Arita. Japan was modernized first with Western technological inputs and mass production management knowhow. The newly-emerged global center of blue and white porcelain production eventually overtook the birthplace of this type of porcelain in technological sophistication and quality control. To a certain extent, that is still the case today. This chapter looks at how the export or trade ceramics were fired in Arita and Jingdezhen based on textual analysis, historiographical work and observation studies for Arita porcelains. Continuing from the previous historical survey, in this chapter, I am interested to examine three issues. First, how modernity has innovated new designs in Jingdezhen and Arita beyond traditional stylized aesthetics. Second, the chapter will look at how energy use has been transformed with the progressive use of coal and then diesel, gas-fired and electric kilns. Third, I will look at how entire ecologies are built up around the creative and production systems in Arita and Jingdezhen.

METHODOLOGY

The methodology for this chapter is based on observation studies and interpretive work on textual and visual materials for analysis. The analytical perspectives are eclectic but centered on fieldwork in Fukuoka Kyushu Japan in order to capture the imagination, ideas and utilitarian use of fired clay (specifically blue and white Arita porcelain) found in the average Japanese household and commercial spaces. How these objects are produced, fired, used and collected is the main focus of this chapter. I am analyzing blue and white porcelains from Arita by applying an eclectic set of theories drawn from critical art criticisms, history of energy transitions and the social history of ceramics use. In my fieldwork, I limited the visual and observation study to only blue and white wares, particularly from the Arita region. The reason for doing this is because blue and white wares are perennial favorites of porcelain consumers and collectors. They are found in households and commercial establishments in large numbers. They are easy to spot, especially in the form of broken shards in the ground or in shops and other commercial establishments. Arita porcelains are recognizable with their stylized drawings and form, particularly for traditional designs (the so-called "Imari" style). Many of them have marks found beneath the porcelain wares and some of them have stickers which indicate the "Arita" label. Porcelain from Arita (known as Imari before the late nineteenth century) are also extensively used in Japan and exported overseas. I also conducted my fieldwork in Arita as well as Fukuoka in the Kyushu region. While the distribution of Arita wares is nationwide, Arita blue and white wares can be found in the immediate macro-region in which Arita is located and are probably endearing to Japanese people and consumers who live in the region of Kyushu since it is an internationally-recognized symbolism of Saga Prefecture and Kyushu region itself. Perhaps, because of the factor of proximity, I was able to locate many Arita ware collectors easily. In Hong Kong where I was based at the point of writing this chapter, I have attended presentations by visiting Saga exchange student delegation and also collected tourist brochures from Saga prefectures and pictures as well as explanations of Arita (or Imari) which clearly indicated Arita as an important historical narrative in that region. Therefore, I chose Kyushu's Arita and Fukuoka areas for my fieldwork. While the focus of this chapter is on Arita blue and white porcelains, it is not possible to discuss blue and white porcelains without mentioning Jingdezhen in China, the original historical area in which they are first mass-produced. Therefore, Jingdezhen serves as an important comparative case study for this chapter.

The unifying theme of the three main components of this chapter (mechanism for design innovation; creative ecology and community; energy use and production) lies in the fact that Arita is a "total" concept, with an autonomous existence that incorporates the function of manufacturing (production), distribution (retail and sales) and technological advancement (design and kiln innovations). Arita ceramics production is not possible without highly-specialized segmentation of ceramics-making functions. As a living ecology of individuals (consumers, artisans, retailers) and organizational structures (kilns, studios, shops and museums), components of this ecological system constantly interact with each other to educate, explain and enlighten consumers (in Japan and overseas) about the aesthetic beauty of this form of fired clay and fulfill the utilitarian needs of consumers in households and restaurants. The Arita producers, distributors and retailers efficiently reach out to other parts of Kyushu, including Fukuoka, the largest city in that region, where I did my fieldwork. Arita wares were distributed to suburban malls, city museums and restaurants. From regional distribution, producers start to spawn national and international distribution network, eventually producing or sub-contracting Arita-style wares overseas. The last step of overseas production represents the ultimate globalization of Arita porcelains, where porcelain manufacturing facilities no longer need to be located in Japan and are subcontracted to kilns in China and Thailand. With its retailing, production and museum spaces, Arita as a "total" concept resolved a common challenge faced by artists and their communities. One complaint that I received from my fieldwork in ceramics-making locations elsewhere (locations that are not specialized ceramics-making regions or towns like Arita or Jingdezhen) is the lack of support from the community for their artwork production. In other artistic studios or clusters that I studied in Kagoshima, Hong Kong and Northern Kyushu, production is dispersed (not segmented and specialized like Arita and Jingdezhen), trained manpower is scarce (e.g., specialized labor like *katamaki* or kiln hands) and infrastructure is lacking (including marketing, museum, kiln and retail infrastructure). They lack the scale, community support and infrastructure and identity of a porcelain-making town like Arita or Jingdezhen for the construction of a comprehensive creative and production ecological system. These infrastructures took centuries to build, accumulate human and financial capital and gain legitimacy as an internationally-recognized ceramics production center.

LITERATURE REVIEW

Three categories of existing literature are reviewed for this chapter. The first set of literature is related to how Japan has continually and successfully indigenized features from traditional Chinese culture, designs and pre-modern technologies to construct its own indigenous culture, e.g., historically indigenizing Jingdezhen designs before developing unique designs. This spirit of indigenization is then extended to the process of learning from Western technologies and tapping into the West's manufacturing ideas for the modernization of Arita porcelain production. Through this East-West hybrid combination both in production techniques and designs, Arita as a kiln, a design center and a ceramic art manufacturing site came into being. Continuous innovation remains a hallmark for Arita ceramics-making industry today. The second set of literature is related to energy use. I am arguing that indigenization cannot be possible without the help of improved production technologies. In the case of Arita, mass manufacturing ideologies, technologies and techniques made the mass production of modernized Western-style utilitarian wares possible. The production of large numbers of wares with roughly the same specification requires more efficient sources of fuels but, at the same time, the process generates polluting emissions. The rise of environmental consciousness influenced fuel use and kiln technologies to evolve and change. The third set of literature is related to production ecology. Without an entire ecosystem to produce, market, design, fire, distribute, display and sell Arita wares, it would not be possible for Arita porcelains to enjoy its status as a leading porcelain producer in the world, with indigenized traditional designs or cutting-edge creative designs.

MODERNITY AND DESIGN INNOVATION

Japan's attempts at indigenizing historical influences from continental East Asia and then the West after the Meiji Restoration 1868 is well-known to students and scholars studying Japan. However, surprisingly, when I tried to search for theoretical literature related to the idea of "indigenization," whether in terms of technology or design, I was unable to locate such works easily. The critical research questions that I had in my mind were whether "indigenization" is a form of ideology, philosophy, applied methodology, value system or developed theory. As I went through the existing literature related to this topic, I realized that it was not any of the above but, rather, an eclectic collection of wide-ranging ideas that were not necessarily coherent or deterministic. It is an evolving discourse which complements Japan's culture of continuous improvements and constant innovation. In my own review of existing literature, the most relevant and important writing in this aspect, though not directly related to the idea of indigenization, is Jan Nederveen Pieterse's interpretation of cultural exchanges. In his article "Global Multiculture: Cultures, Transnational Culture, Deep Culture," there are three points which are relevant to the current chapter. First, he argues that technology (especially digital technology) is now transnational and crosses boundaries to cross-pollinate with different cultures.[3] While trade porcelain exports were incomparable to digital technologies in terms of reach and extensiveness, they were historical agents of cultural dissemination, often transferring ideas to other cultures which were then indigenized by those cultures. In the seventeenth century, Imari (now Arita) designers may have received influence from Jingdezhen in terms of early blue and white designs but the transnational transfer of design ideas did not remain unidirectional statically. The consequent multi-directional flows of ideas and artisanal interactions resulted in the emergence of indigenous unique Japanese designs. Meissen in Germany, Royal Copenhagen in Denmark, Delft in Holland, the Annamese ceramicists, the Okinawan porcelain makers, Spode and Wedgewood in the UK eventually adapted their own blue and white designs from their interpretations of Jingdezhen and/or Arita designs. These cultures received the blue and white porcelain art soon became distinguishable and unrecognizable from original Jingdezhen and/or Arita designs while adaptations and indigenization became innovation.

Second, Pieterse defines culture as an intangible element that is not static and constantly evolving: "culture in a general sense is human software and know-how, or all that is learned in the process of being human. This learning is open ended, never complete, always in flux, and not necessarily bound to a particular location."[4] This contextual reading of "culture" is an important argument in this chapter. Embedded in this case study, I am arguing that Arita kiln's absorption, indigenization and application of blue and white porcelain-making culture from continental Asia (China and Korea) were not uncritically accepted and reproduced. There was dynamic reception on the part of Japanese potters and ceramicists who used their knowledge to figure out how to continually improve, innovate and create new designs. When

Western technologies and knowledge became available, they were incorporated into the East Asian knowledge base, producing new hybrid cultures that did not entirely imitate those of the original source of influence and are distinguishable from the new elements that were incorporated into traditional arts. The hybrid cultures became autonomous entities with their own existence. The principles of artistic creation applied in that process of indigenization becomes rooted, established and entrenched amongst communities of producers, collectors, traders, consumers and scholars who study them.

Another argument raised in Pieterse's writing is only coming into being in the contemporary setting. Historically, blue-and-white porcelains whether from Arita or Jingdezhen are not yet part of a "global multiculture"[5] since they do not incorporate multi-ethnic components and neither are they sufficiently globally distributed to qualify for this terminology. But they do show "vernacular interests," "nativism" and "local software" characteristics[6] when blue and white porcelain begins to be manufactured in different parts of the globe outside Arita and Jingdezhen in locations such as Delft, Meissen, Copenhagen or the US where they are designed and produced locally with expressions of vernacular sensitivities. These particularistic local interpretations of original blue and white porcelain designs from Jingdezhen, artistic expressions and consumption trends then feed into a superstructure of global consumption of blue and white porcelain. While the global production and consumption of porcelains are outside the scope of this chapter (which focuses mainly on Arita and Jingdezhen varieties of porcelains), its distribution worldwide, even in the pre-modern period to many parts of the known world, demonstrates the extensiveness of blue and white porcelain consumption. In China and Japan alone (and elsewhere), the appearance of blue and white designs is almost synonymously associated with the idea and concept of the physical material of porcelain itself.

Historically, Arita (and the older terminology Imari) showcased Japan's porcelain-making skills in producing utilitarian but highly attractive and aesthetic wares for international consumers. Innovation drove the Japanese to create new designs that distinguished Arita blue and white porcelain works from traditional stylized Jingdezhen designs. While objects and motifs like dragons and flowers remain popular designs, they are indigenized through local Japanese design interpretations.

In my fieldwork in Japan, I specifically looked out for Imari-style porcelains to support the idea that Japanese indigenous designs were not something new but that the process of indigenization was constantly occurring in Japanese interactions with other cultures and to accommodate consumer trends, modern and contemporary usage of porcelains as well as customary or/and religious functions of utilitarian ceramics (please refer to figure 4.1 for an example of a ceramic ware that incorporates functionality, traditional *sometsuke* (blue and white) aesthetics and customary/religious usage).

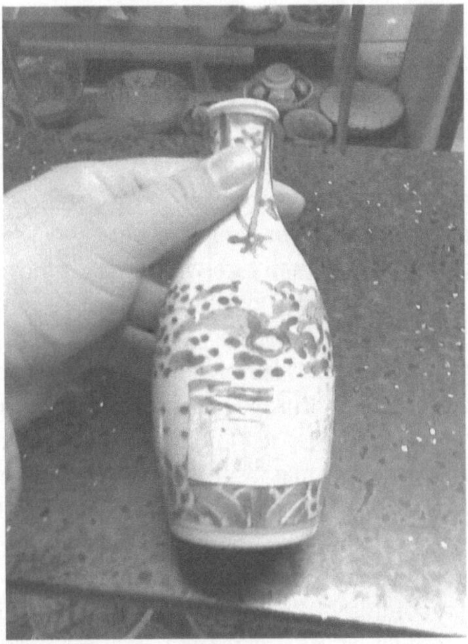

Figure 4.1. The streamlined shape of the bottle with a thin neck and heavier body is drawn with uniquely Japanese designs for phoenixes (Photo taken by author). Besides the drawings on the porcelain, design innovation sometimes applies to the shape of the vessel itself. The bottle in photo 1 which I found in an art gallery in Fukuoka during my fieldwork in Japan in 2013 carries water for offering to Shinto gods in small altars. It is a blue and white bottle with thin body done in the Imari style, likely to date back to the late nineteenth or early twentieth century. The streamlined shape of the bottle with a thin neck and heavier body is drawn with uniquely Japanese designs for phoenixes. The narrow neck and wider body is probably designed to accommodate the cut stem of a plant offering at the altar. According to its veteran collector-owner, it was used to hold water and plant offering to Shinto gods.

LITERATURE REVIEW – WESTERN INFLUENCES

In my fieldwork in Japan in 2013, I visited a folk ware showroom which featured local folk arts and crafts (*mingei*) of Northern Fukuoka Prefecture Kyushu. The ceramic bottle on display below was an exemplary example of a soy sauce bottle displayed at the Paris Exposition in 1900, used to bottle Yoshida Company's soy sauce brew. The high-quality bone china porcelain for a liquid that is ubiquitous with East Asian cooking must have drawn curious attention from visitors with its high quality aesthetic painting (finely-drawn for a common utilitarian mass-produced ware). From a Japan-centered perspective, it must have complemented the traditional features of the Japanese booth at the Exposition which also featured a traditional Japanese garden. Another reason for featuring this bottle in this chapter is because of European tastes for Imari wares in that particular synchronic period of history. A full detailed discussion about French and European aesthetic tastes for Imari wares in the seventeenth and eighteenth centuries are outside the scope of this chapter but readers can refer to Imai Yuko's paper "Changes in French Tastes for Japanese Ceramics" for more detailed discussion in this area.[7] Western European tastes for Japanese blue and white porcelains and other traditional art forms eventually grew into an art movement. Japonisme is a form of hybridized art made possible because of renewed interactions between Japan and the West in the early modern period. Materially, it resulted from a fusion of Japan and European art-forms when traditional Japanese crafts were interpreted by European artists and artisans who also indigenized traditional Japanese artistic techniques to create innovative art-forms. Many of these crafts and craft-making processes considered cutting-edge or different by Western artists were commonplace handicrafts in Japan. The process elevated common crafts to a higher level of avant-garde artistic production. The autonomy of the japonisme movement is well-discussed in Alicia Volk's important volume *In Pursuit of Universalism* which argued that an art form acquires its own set of meanings and symbolisms after being exposed to interpretations and re-interpretations by scholars, critics, artists and other stakeholders in the ceramics art world at a particular synchronic point of historical time.[8] While detailed studies of the japonisme movement and European aesthetic tastes in detail are outside the purview of this chapter, these two ideas are important for explaining how Japanese exports to the West caters to a Eurocentric preference for traditional artistic and artisanal wares. Utilitarian wares displayed at the Paris Exposition such as the soy sauce ceramic bottle were drawn with Japanese designs but have textual materials written in English on its surface. The explanation signboard for the Imari soy sauce bottle in the folk art showroom indicates it is a gift of the seventeenth generation of Yoshida-ya (House of Yoshida,[9] the soy sauce maker) to the *mingei* showroom.

In the same folk ware showroom, I was served Japanese green tea in blue and white ceramics although their provenance is unknown. The selection of blue and white wares may be due to aesthetic reasons since it fitted with the blue theme of the showroom which displayed dyed blue traditional fabric. This Northern Fukuoka *mingei* non-profit showroom that was run by local authorities therefore highlighted the blue and white theme that unified some of the art-forms found in the region. Away from the *mingei* showroom and within the same district, I also searched for traces of blue and white wares in old historical shop-fronts in a folk arts-producing district in northern Fukuoka. In my fieldwork, I discovered at least two shop-front displays that showcased either Arita or Ko-Imari-ish designs.

In the absence of existing literature on the idea of indigenization, I proceeded to characterize Arita's process of "indigenization" in the following three ways. First, at the broadest level, it is a form of East-West hybridity. Arita porcelains (and Imari in the past) combined porcelain design influence from continental Northeast Asia with Japan's interpretation of modernity and the use of Western technologies and technical knowhow (I will discuss more about the technological factor in the section on kilns and energy use below). It adapts and indigenizes traditional designs from China for example and tailor-made these products for Western European customers while adding its own design inputs into the process starting from the early pre-modern stage. In the modern and contemporary settings, Japan developed its own unique idiosyncratic styles that were emulated by others. Second, Arita porcelains represent functional adaptation. Traditional porcelain shapes found in Jingdezhen were modified to suit different functions in the household or commercial establishments. Some of these shapes were adaptations of traditional porcelain wares to cater to Western tastes and household usage in the process of exporting them. Third, Arita indigenization is constant, always evolving and non-static. It is reactive to environmental factors, such as fashion trends, fresh innovative ideas about the use of porcelains and the evolution of production technologies. This point correlates with the writings of Jan Nederveen Pieterse in the review above.

IMAGINATIONS AND CONSCIOUSNESS OF BLUE AND WHITE PORCELAINS IN DAILY LIVES

In discussing the existence of blue and white Arita porcelains in the daily lives of Japanese people, I found Goro Shimura's trade publication *The Story of Imari* written from the perspective of a collector useful. There are three relevant perspectives he highlights in the volume that is useful for this chapter in the discussions about functional wares used in the daily lives of Japanese people. First, he points out the influence of functional use in the creation and development of Arita porcelains in Chapter Two by tracing the influence and impact of tea culture[10] (perhaps one of the oldest sources of ritualized influence on the functionality of Japanese ceramics) on Arita wares and ceramics art. The Japanese tea ceremony constructs highly ritualized movements and prescribes rules for consuming and appreciating the tea and the tea wares that hold and contain the tea brew. This is probably one of the earliest examples of functional rules governing the design of porcelain wares which have to accommodate these ritualistic rules. Other than the important set of ritualistic rules governing the Japanese tea ceremony developed by Sen no Rikyū that governed the size, shape and design of Arita tea wares, some of the priceless porcelains made by Arita and other ceramics makers served as important tributary gifts for diplomatic exchanges with alliance partners and political rivals.[11] Outside these elite socioeconomic class activities, Shimura's volume also features a wide array of covered cups, lidded wares and functional wares like the *danjū* (stacked-up tiered containers with a lid on top) designed to hold different types of food for special occasion like the New Year[12] that were used by ordinary folks. There is, however, a limit to interpretations of functional designs in some of the pieces that Shimura featured in his book. Sometimes, aside from functional reasons, designs are based on the need to accommodate aesthetic presentation of textual materials such as the fan-shaped dish featured on p. 109 of Shimura's publication,[13] which aligned panels of different designs in stylized frames together to make up a coherent story or design concept. As Shimura is a collector of old Arita pieces (specifically Imari wares), his materials on functionality is limited to wares dating back to that time period. Functional designs continue to be relevant and important in current designs and settings. In my observation fieldwork, I spotted some interesting contemporary Arita pieces which embody both innovative designs and functionality.

CERAMICS ECOLOGY AND COMMUNITY

In the contemporary setting, the public consciousness of blue and white as representative porcelain designs remains strong amongst consumers, collectors and museum visitors in Japan and China. I have had chance to do some fieldwork observation in the so-called "Greater China" region of Hong Kong and Taiwan for this purpose. In Japan, I also found public displays of both Jingdezhen and Arita blue and white wares. The locations where they are found are eclectic, in restaurants and shop-fronts and other locations associated with the use of porcelains. In my fieldwork in Japan, I also observed an Arita ware sale bazaar in a Fukuoka suburban mall. This fair featured hand-drawn Arita pieces sold at discounted prices.

Besides shop-fronts, I also observed Japanese household use of Arita pieces during my fieldwork. Consumption is an important process that drives fashion trends, pricing, production techniques and manufacturing philosophies and ideologies. Part of the ecology of Arita ware usage lies in the daily consumption of its utilitarian products. In the location of Fukuoka where I carried out my observation, the marketing and sale of Arita wares was extensive and could be found in malls, households and shop-fronts in my observation studies. The mall locations were the terminal ends of an entire ecology serving popular consumption. Consumption provides the revenues necessary for the production, design and innovation of Arita products. Funds are then recycled back to the producers in the consumption superstructure. Besides the mall, there were other less commercialized outlets which showcased

Arita wares for collectors who want more selective pieces that are not mass-produced for regular mall shoppers but for the informed museum visitor. During my fieldwork at an avant garde downtown Museum featuring mainly Asian Art in Fukuoka, I recorded visuals of exclusive Arita wares on sale and display at the museum shop with very simple functional lines reminiscent of diner sets found in North America.

ENERGY USE AND PRECISION-FIRING

The ecological system of a porcelain town like Arita is incomplete with only consumption activities, the production system must be equally specialized and integrated with resources found in that area. The origins of proto-industrial production of blue and white ceramics started in Jingdezhen during the Mongol-ruled Yuan dynasty where ceramicists were systematically organized by the state into production houses for making large number of wares. Division of labor needed for mass manufacturing also expanded during this period. These techniques eventually migrated to Korea and Japan and were adapted for Arita's own mass manufacturing needs. Part Three (or the last three chapters) of Margeret Medley's important volume *The Chinese Potter* is a highly-relevant and useful reading related to the formation of pre-modern industries in Jingdezhen for manufacturing blue and white porcelains. She makes three interesting arguments here. First, Medley boldly argues that the under-glaze technique was the "last great innovation" in porcelain-making history.[14] This basically refers to the use of cobalt blue oxide in drawing and then protected by a layer of glaze applied over it, locking the colors into the porcelain surface and also creating a strong protective layer when fired up to high temperatures. The protective layer and firing makes it possible for blue and white shards to be identified easily, given the vibrant blue color standing out against a white base, and also protects the cobalt-based brushstroke drawings against the ravages of time and corrosion by seawater. The second major argument in Medley's Part Three section points out the adaptation of Jingdezhen's industrial machine to prevailing fashionable tastes and trends in the European markets. They produced wares that the customers needed, effectively commercializing the product of blue and white porcelains.[15] It is this quality that defined Jingdezhen and later Arita trade ceramics export industries, enabling the product to go global effectively. It was not only an artistic ware but blue and white porcelains were also commercial products with functional and utilitarian purposes. Third, perhaps Medley's most important point of all, was that Jingdezhen underwent an industrial revolution with division of labor, manpower skills training, the hiring of specialist part-time laborers complementing production needs and also set up a capitalistic financing system to run the kilns.[16] The kilns used during this period were "multiple some-vent, down-draught type, known as bee-hive kilns....previously had been developed in Honan."[17] A typical wood-fired kiln in Jingdezhen is domed-shaped with a chimney and timber pieces stacked up by the side of the kiln body for feeding into the firing chamber. The base is made of stone bricks and the covering for the dome made of pressed mud. Wares are stacked up inside. These were pre-modern technologies that required large numbers of manpower to adjust the air chamber and quantity of timber fuel inputted into the body of the kiln. They also had higher failure and accident rates due to the manual nature of temperature control and kilns that operated on the labor-intensive manual insertion of timber took a longer time to fire, sometimes over days. Several rotating teams must also be on standby for this process. In imperial kilns, minor defects meant entire batches of porcelain had to be destroyed to keep only the most pristine pieces. The labor-intensive nature of the firing process is mentioned in Tolliday and Yonemitsu's paper who pointed out that Arita built its industrialization on brute labor which depended on hands-on laborious work, such as the *kamataki* specialists who were on standby to feed the wood-fired kilns with timber and had their noses and faces perpetually burnt.[18] From a productivity point of view, it was wasteful, labor-intensive and, as environmental consciousness grew, environmentally-unfriendly as well.

After modernization, the Japanese introduced improved square kilns to Taiwan, typically built with a brick chimney, air hole openings by the side of the kiln and a dome that is square in shape. Coal instead of timber is used as energy fuel, allowing the kiln to produce greater output of porcelains and ceramics. Coal-fired kilns represented advancement over traditional climbing (*noborigama*) and snake kilns in Taiwan. Gradually, newer technologies like roller trays, gas and electric kilns were introduced to Taiwan, completing the energy revolution in ceramics production. Contemporary gas and electric kilns had significantly lower emissions compared with timber or coal. They resolved the problem of random movement of air, ash and fire within the kiln, allowing for ultimate precision firing although ceramicists complain about highly predictable results. As industrial kilns, however, they facilitated precision firing for mass produced Western-style utilitarian wares. There are also fewer defects.

Precision firing makes the production of thin, standardized utilitarian porcelain wares possible. Japan probably reached a leading position in marketing, distribution and retail sales of high quality bone China wares in the postwar period of the twentieth century. Arita porcelains, including bone China fine porcelains, continue to be highly sought-after by consumers acquiring utilitarian ware. However, in terms of branding, Arita wares are preceded by Western-style fine bone china specialist makers in Japan such as the likes of Noritake (an internationally-recognized brand), Fukagawa (a purveyor of fine porcelain for the imperial household), Kōransha, etc.[19]

These specialized bone china fine porcelain makers successfully captured a niche high-end market of the industry. Arita kilns, on the other hand, cater to a pluralistic range of porcelain consumers and collectors who do not only focus on high-end fine bone china products but also manufacture affordable daily wares. The informative journal article on the case study of Arita industrialization by Steven Tolliday and Yonemitsu Yasushi indicated that Arita only absorbed Western technology to improve its artisanal craft products but did not participate in manufacturing Western-style utilitarian wares or industrial products.[20] This was Arita kilns' attempt to maintain its own identity in design and functional concepts and not conform completely to producing Western fine bone china wares. Maintaining high quality and displaying traditional Imari designs, some Arita products may not be high-end delicate thin bone china products but are aesthetically pleasing for daily use in households and restaurants. They are also sturdy enough to survive daily use, dishwasher use and daily scrubbing without chipping easily.

Generally, the two representative Western-inspired ceramics maker, Koransha and Fukagawa, produce utilitarian wares that show at least three similarities. There are clear signs of Western influence in the two examples. They have thin and smooth bone china porcelain bodies, characteristic of high-end Western-style porcelain wares. The drawings on both wares are highly stylized and symmetrical with high-quality finishing. These two manufacturers are not the only kilns firing such porcelain wares. Given the export potential of Western-style bone china wares, other lesser-known porcelain manufacturers also stuck to traditional Western-style fine bone china porcelain designs mainly for export to the West and the "Greater China" market. This becomes an interesting case of "Greater China" consumers acquiring Japan-made wares that emulate Western utilitarian porcelains. To a certain extent, Koransha, Noritake and Fukagawa have even overtaken Western porcelain manufacturers in terms of brand recognition and gain the legitimacy of the endorsement of the Japanese imperial household. Outside these major producers, many other Japanese makers also manufacture high-quality bone china wares for export.

Most Arita wares are not confined to only very fine thin bone China wares. One can commonly find Arita wares with stout and thick ceramics bodies and highly-stylized designed but hand-drawn with well-practiced strokes rather than highly-symmetrical embossed prints. These wares can be found in all parts of Japan, in restaurants, households, commercial outlets and a variety of commercial locations.

The production of high-quality fine bone china wares are made possible only with advanced kiln technologies that facilitate precision firing. According to Tolliday and Yonemitsu, the revolution in energy fuel use came in the 1960s:

> Producers began to move from wood-fired to coal-fired to oil-fired kilns, which had higher rates of successful firing and which could be used twice as frequently (though many of the new kilns were very polluting). In the late 1960s and early 1970s, gas and electric kilns came in which were cleaner and even more effective. While only 50–60 per cent of ware came out successfully from wood-fired kilns, gas-fired kilns averaged 95–97 per cent. Wood fueled kilns could only be fired two or three times per month while gas-fired shuttle kilns could be fired 10–12 times per month and gas-fired tunnel kilns daily.[21]

There are several implications of this evolution of fuel use. First, fuels became more efficient with increasingly less accidental rates. Wastage therefore is minimized but the challenge was that coal and oil (coming into use in the 1960s) led to more pollution which was resolved only when gas and electric kilns came into being. Because gas and electric kilns required less manpower to operate, more individuals could now engage in ceramics-making activities on a part-time or full-time basis and could fire their ceramics more frequently. The development of these mobile and easy-to-use kilns was fortunate because, as labor costs went up and the increase in yen value, the practice of using *kamataki* (specialized kiln hands) became less feasible so one-man-operation (OMO) automation became a necessary solution to reduce production costs. This stage represented a milestone from the production side. Collectors and consumers worldwide have access to affordable Arita porcelains after proto-industrial production came into being during the Yuan dynasty period and spread to Japan and mass manufacturing techniques in the modern period made ceramic products even cheaper after Japan learnt these techniques from the West. But the fuel evolution and kiln revolution now made ceramic production accessible to amateurs and professionals alike. This development placed production means within reach to many more craftsmen or would-be artisans. Because of technological improvements, both production and consumption capabilities have equally proliferated to a larger section of socioeconomic classes.

CONCLUDING SECTION AND LIMITATIONS

It is difficult to quantify with precision the use of blue and white porcelains in the Japanese people's daily lives. Despite the lack of statistics, the ease in which the author was able to locate Arita-produced or Arita-design wares in shops, restaurants and households is indicative of the ubiquitous nature of Arita wares in Fukuoka and Kyushu. The success of Arita porcelains have spawned imitations where Arita-like porcelains are now sold within and outside Japan, catering to consumers who are keen to own and use Arita-style designed utilitarian wares but not necessarily willing to pay high end prices for them. Imitations and local adaptations of Arita designs is likely to represent another phase in its ceramics production history as the site of production itself no longer needs to be situated in Japan. Arita as a terminology may have made the migration from the name of a locale to an intangible concept itself. Arita wares may no longer need to be defined strictly in accordance with the location in which these wares are made and instead becomes as signifier and symbol of an artistic movement of style in ceramics, much as japonisme had been in the distant past as a general art form. For majority of the consumers, authenticity is less important than the quality of the ware pegged at a reasonable price (collectors are probably exceptional to this). In my fieldwork visit to the discount 100 yen stores in Fukuoka, a large number of the inventory were occupied by "Arita-style (*Arita-fuu*)" wares made in other Asian countries (please refer to figure 4.2).

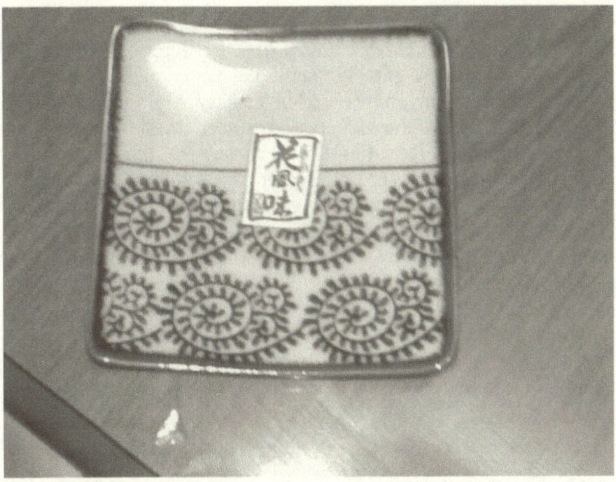

Figure 4.2. Arita-style wares made in Thailand or China found in the 100 yen shops in Japan during my fieldwork (Photo taken by author). The photo shows one of the Arita-style wares that I collected in my fieldwork in the 100 yen shops in Japan. Most of them are made in China or Thailand. For the non-specialist, it is challenging for observers to tell apart the difference between Arita-style and Arita-produced wares. It is a testimony to Arita wares' popularity with overseas consumers.

From the production point of view, the development represents a logical continuation of the process that was started in the 1960s when Japanese wages started to climb up and the yen gradually strengthened, making the manufacturing of less value-added products unfeasible. Like other similar price-sensitive industries, they were shipped out to other Asian economies. From my collection of "Arita-style" wares not made in Japan, two countries, China and Thailand stand out as prominent alternative manufacturing centers outside Japan for such wares. Ironically, the production of imitation Arita or Arita-style wares may actually ensure the survival of the Arita artistic style and also the proliferation of related wares to the rest of the world.

NOTES

1. This chapter is funded by the Hop Wai Short-term Research Grant from the Chinese University of Hong Kong, titled "A Comparative Study of Energy Use in Jingdezhen and Arita Porcelain-Making Regions: A Historical Overview and Historiography of Kiln Modernization and Aesthetic Innovation of Export Wares."

2. *Qinghua* refers to the blue-color designs on the white based of ceramics originating from Jingdezhen while *sometsuke* refers to the blue indiglo color of traditionally-dyed Japanese fabric.

3. Jan Nederveen Pieterse, "Global Multiculture: Cultures, Transnational Culture, Deep Culture," in *Hybrids, Differences, Visions: On the Study of Culture*, ed. Claudio Baraldi, Andrea Borsari, and Augusto Carli (Aurora, CO: Davies Group, 2010), 1–2, http://janneder-veenpieterse.com/pdf/Global_multi.pdf.

4. Ibid., 2.

5. The subject of this chapter: Ibid.

6. Ibid., 10–11.

7. Imai, "Changes in French Tastes for Japanese Ceramics."

8. Refer to: Alicia Volk, *In Pursuit of Universalism* (California: UC Press, 2010).

9. Through an online search, a possible surviving entity of this company could be the Yoshida Shōyu Soy Sauce company, now based in Yamaguchi prefecture that was established in January 1897 (or thirtieth year of the Meiji period). Given that it is a family business and that the seventeenth generation is still identifiable, I speculated that the company is still surviving although management may have changed or the company itself restructured and reconfigurated. If it is indeed the Yamaguchi-based company, information about this company can be found in: Yamaguchi Prefecture Shunan Industrial Company Database, "Yoshida-ya shōyu," 2011, accessed February 15, 2013, http://www.shunan-marketing.jp/detail.php?cci=1&manage=2412; Ameba Japan, "Yoshida-ya Shōyu," February 5, 2011, accessed February 15, 2013, http://ameblo.jp/yoshidaya-soysauce/. This is the only surviving Yoshida soy sauce manufacturer, the other is a contemporary company that distributes to the US market.

10. Goro Shimura, *The Story of Imari : The Symbols and Mysteries of Antique Japanese Porcelain* (Berkeley California: Ten Speed Press, 2008), 13–40.

11. Ibid., 14.

12. Ibid., 103.

13. Ibid., 109.

14. Margaret Medley, *The Chinese Potter: A Practical History of Chinese Ceramics* (London and New York: Phaidon Press, 2006), 176.

15. Ibid., 216–17.

16. Ibid., 171.

17. Ibid., 171.

18. Tolliday and Yonemitsu, "Microfirms and Industrial Districts in Japan," 49–50.

19. While Kōransha and Fukagawa are located in Arita, they are typically autonomously classified as an independent genre of its own in the porcelain typology. Producing stylized thin-body fine bone china porcelain, it has a distinctive style of its own with strong Western influence that is distinguishable from the typical Arita wares.

20. Tolliday and Yonemitsu, "Microfirms and Industrial Districts in Japan," 47.

21. Ibid., 51.

Chapter Five

Ideas about Resource Use and the Natural Environment in Pottery-making

A Historical Comparative Case Study of Two Communities in Hong Kong and Japan[1]

This chapter has its origins in a grant application. In applying for the grant, I was interested to examine contemporary ceramic artists' ideas about environmental conservation and energy use in their line of work. There are three important questions which I wanted to address in this chapter: First, do professional potters consciously think about environmental conservation and energy efficiency in their professional work? Second, if these factors are considered, how do they feature into the potters' production and work processes? Third, what are the justifications behind including environmental factors and ideas about energy use in the professional potters' work processes? This chapter is a comparative contrast to the previous preceding chapters in the following ways. It focuses on pottery-making artistic clusters that are located in densely-populated urbanized locations instead of a single town dedicated to making potteries like Arita, Jingdezhen or the case study of Yingge which comes later in the book. It looks at how choices made in selecting kiln technologies are also shaped by the external environment. Most importantly, the Hong Kong case study of disparate pottery-making artistic clusters and industrial park villages show a different kind of configuration in pottery-making ecology that serves as a contrast to the other three case studies covered in this volume which are dedicated one-stop ceramics-making towns with strong institutionalized features to support the industry.

In this chapter, I will not engage in a definitional debate about the different classifications of potters and their art because it is not my main concern here. I am also not studying industrial potters with large-scale production or completely automated processes of pottery-making. Instead, I am interested in small communities of potters who rely on their own means to source for materials and produce unique pieces that are acquired by collectors and consumers. It does not matter to me whether the potters are defined strictly as folk or contemporary artists. A unifying element that I wish to employ here is brought up by my Hong Kong informant Ms. Florence Yim:

> There has been only a certain nos. of ceramist persisting on hand-made studio pottery…There has been local potters being influenced by Bernard Leach, Shoji Hamada, etc. since the 60's, but most of them had not treated ceramic making as a kind of 'folk' heritage (old and vintage), but stressed more on its 'hand-made' and/or 'expressive' qualities instead.

Several qualities are mentioned here that are of interest to me: the qualities of hand-made products; local and international influences on Hong Kong-based potters and the expression of traditions and innovations. These are the defining features which I have also utilized in picking out the Brian Moeran's study of the Sarayama community and Hong Kong's fourth-generation ceramicists as my case studies. I have also included the Hong Kong informants' information about their observations of the Shigaraki potter community for reference. My HK informants and the Sarayama potters in Moeran's study are characterized by small-scale operations, reliance on a small community of like-minded individuals and they produce artistic pieces. Their works are the outcome of conscious selection of ingredients and materials used for the firing process to achieve certain artistic goals compared with the Taylorist mass production of large-scale pottery-makers. For the latter, profits, productivity and efficiency are likely to be far important priorities than individual uniqueness and deep contemplation of environmental sustainability or ethical resource use.

From my fieldwork interviews with the Hong Kong contemporary potters carried out on 31 August 2012, I can detect some correlations between them and the *mingei* folk potters in Japan in terms of self-identities and outputs of their craft. I also visited their studios on 6 September 2012. I will summarize their similarities in three main areas. First, there is a conscious element of self-reliance when it comes to sourcing for ingredients for their work processes. While the methods in materials acquisitions may differ, the decision-making processes of the potters show similarities. Both Sarayama potters in Moeran's study and the Hong Kong potters that I interviewed rely on conscious searches for the best possible ingredients for their crafts. In making these choices, the criteria of pricing, quality and natural ingredients comes into play. Related to the subject matter of this chapter, environmental sustainability of resource use and costs of energy are also major considerations here. Secondly, both sides have accumulated East-West influences in terms of ideas about modernity. Both communities of potters face clear challenges and transformations brought about by modernity and the technologies that they represent. Those that transformed the Sarayama (Onta) pottery community in Moeran's study are rather similar to narratives about modernity and technology that the Hong Kong contemporary potters have described to me:

> [Referring to Onta] Some of the techniques of production (like the use of water-powered clay crushers) are no longer, or for one reason or another never have been used in these communities; but all of them have made use of modern technological developments to increase production' all of the have had to learn to cater to a predominantly urban market…[2]

From the passage above, the main concerns with modernity and its associated technologies are output productivity, the relentless march of technology in replacing manual labor and the powerful influence of urban consumption. I will examine this issue of modern technologies and their impact on the environment and energy use in greater details in the main text. The third area in which there are similarities between the Sarayama and HK potters is found in functional ideas about energy use based on economics. The cost-effectiveness of using energy resources and fuels constitutes the core considerations of both communities of potters when making decisions about energy use and conservation. I will also examine this in greater detail within the main text.

While there are similarities between the Sarayama folk potters and the Hong Kong contemporary potters, there are also important differences and contrasting features as well. First, Moeran's subjects have been making pottery for over 250 years[3] and so, there is an element of time-tested traditionalism in the way they made their crafts. According to my oral interviews with the fourth generation of potters in Hong Kong, the current lineage of professional potters in Hong Kong is traceable back to the 1940s. The difference in the histories of pottery-making traditions in the two locations has implications for the environment and energy use. Entrenched ideas about production make it more difficult for the Sarayama potters to reconfigure or reform work processes compared with the individualistic operations of the Hong Kong contemporary potters. In Sarayama, the persistent use of traditional multi-generational clay crushers and the timber-fired climbing kilns are examples of emphasis on keeping traditions and culture alive.[4] Lo's experience in Shigaraki corroborated this point. He noted climbing kilns tend to face the energy-efficiency disadvantage of having differential heat intensities found at various sections of the kiln chambers. He mentioned that some Western innovators reversed the incline of the sloped kilns so that it is fired from top to bottom and the wind draft's downward direction would distribute heat more evenly and conserve energy fuel use. But he also noted that the traditional potters in Japan rejected this technology in favor of bottom-up firing traditional climbing kiln, citing mostly preservation of traditions for this decision. He admitted that there could also be an aesthetic effect of differential intensities of firing in a tradition bottom-up approach in operating climbing kilns, producing different types of glaze effects due to the differences in temperatures.

Moeran's Sarayama community continues to observe age-old traditions in almost all stages of the process, including sourcing for natural ingredients. Ideas about the rural environment from which Sarayama derive their natural ingredients to fire their pottery places stronger pressures on maintaining sustainability of those resources in comparison with the Hong Kong potters who are more concerned with the ethical nature of their ingredient purchases. Both aspects, however, are unified in their conscious awareness of environmental priorities. In contrast, persistence in traditional ways of doing things does not hold to the same extent in my Hong Kong case study. Philosophically, the Hong Kong potters' concept of time is rational and utilitarian. My informant Florence Yim, a fourth generation contemporary ceramicist in Hong Kong explained:

> My point of view on folk pottery is merely sharing…I guess the 'folk' heritage which looks old and vintage in nowadays was once 'contemporary' in their times; so, what we are doing now to reflect the spirit of our time might become a kind of 'folk' in future if some kind of technique and/or style are going to be inherited and then passed on and on. Hey, so we should make something good so that we have something worthwhile for passing on, shouldn't we?

Yim did not attribute the goal of pottery-making to any ideological or academic discourses but conceptualized the activity as merely sharing the fruits of her labor with others. She also rationalized the concept of folk art by referring to the relative nature of time. Traditionalism and old heritage were not timeless concepts but are relative to the point of reference used. Overall, to her, pottery-making activity is reflective of rational behavior in creating an art object for instant gratification and reflective of current priorities. And if it is good quality enough, the same object can then be passed on in perpetuity. Yim also believed that a globalized world no longer necessitated the Hong Kong contemporary potters to stick only to traditional designs or ways of making pottery. She conceptualized Hong Kong as a cultural melting pot, accepting influences from the East and West while foregrounding the unique Lingnan culture. Unlike their Sarayama counterparts, Yim and her fourth generation compatriots have all made the switch to electric kilns despite having previous experience in building their own traditional kilns. They made the change after rationally assessing the situation and felt that it took too much work to maintain and fire the traditional kiln.

Besides production methodology, Yim and her compatriots are still exploring the possibilities of extracting their own clay from Hong Kong sources (discussed in more details in the main text) and instead rely on external acquisitions to get their materials and equipment. This option is based on valuable knowledge imparted by the second generation *sifu* (master) Chen Chunguang (or Chan Chung Kong in Wades Giles), famed for his technical knowledge and skills. Yim recalls his advice in this area in 2010:

> Mr. Chan Chung Kong about 2 years ago, there are some sites in HK from which natural resources of ceramic clay or mineral materials for making clay or glazes could be found: (most of them have to be added with other materials to improve the clay body before use): Cha Kwo Ling, Ma On Shan (Pyrrhotite found here), Chek Lap Kok (a quartz mine is found here), Shatin Tung Lo Wan Village (a felspar mine is found here).

According to Yim, Mr. Chan Chung Kong mentioned to Yim in 2010 that he discovered the indigenous clay and mineral sites in the 1960s or 1970s. Yim's views are ambivalent about the presence of local ingredients:

> I've not given up any hopes because finding out local raw materials for ceramic making is an interesting concept, I might take this concept as part of my creative work in future. Yet it's not likely that I would use local material and spend too much time to turn them into workable state in the long-run.

On the one hand, there is a desire and knowledge to use those ingredients which can produce unexpected results and also reinforce the idea of indigenous production but, on the other hand, it may not be feasible in the technical and cost-effectiveness aspects.

Secondly, the artistic environments and consumer markets in Sarayama and Hong Kong have different characteristics and differ in receptions to the potters' outputs. The Hong Kong potters that I interviewed concluded the pottery art-collecting and art-consumption environment in Hong Kong is small compared to what they observed in Japan in locations like Shigaraki. They noted with envy the collegial atmosphere of gallery exhibitions, public support and participation in pottery events in Shigaraki. Chris Lo, one of the fourth generation potters, mentioned that exhibitions were held in casual and warm atmospheres in restaurants in Shigaraki and felt that such casual settings were not possible in the Hong Kong context. In his view, there is conscious cultivation of art awareness amongst the young and the potters enjoyed strong community support. In contrast, they felt that there was a comparatively smaller market for contemporary pottery pieces in Hong Kong (compared to Shigaraki's potter community and compared with the paintings/sculptures collectors' market in Hong Kong) and this posed challenges for their livelihoods and the financial means to sustain their activities. This also has implications for the environment and energy use since the size of the consumer market becomes a natural limitation on the growth of the community and the output they produce. Therefore, resource use is also correspondingly circumscribed.

Thirdly, the collective community organization and cooperative relationships between specialized individuals in Sarayama contrast with the robust individualism of Hong Kong professional potters. In the case of Sarayama, this is summed up by Moeran:

> We have seen that in Sarayama individuals were expected to subordinate their personal interests to those of the household and community to which they belonged. People were expected to live in harmony with others.[5]

My interviewees tell me that the Hong Kong potters do not have the luxury of "ecology" of network individuals specialized in producing outputs at each stage of the work processes necessary for making the final products. The motivations behind pottery-making in Hong Kong are based on personal concerns like turning to pottery-making as a hobby for de-stressing purposes (a form of "meditation"), expression of creativity, authenticity of personal feelings and personal satisfaction. These are individual and personal concerns rather than community priorities. Suggesting more personal reasons for taking up pottery-making, Yim elaborates on this point below:

> The motivation behind pottery making in HK, other than as a kind of art pursuits, is also for educational reason like enhancing fine-motor of children through education. It is also used as a self-healing media in psychological term Chris mentioned on this point when we talked about 'spirituality').

On the other hand, motivations behind pottery-making activities in the Sarayama community include establishing the legitimacy of one's community identity, forging inter-personal relationship for careers and enhancing a sense of community belonging. My informants' experience at Shigaraki corresponds to this reading. They noted that potter positions in Sarayama tended to be hereditary, passing down from fathers to sons. Therefore, there is a strong sense of responsibility to the community in keeping the crafts alive.

The specialized nature of the roles that Sarayama potters and their associates play within the community may be related to the technologies that they used, that is, the use of cooperative timber-fired kilns, as opposed to the Hong Kong potters who relied on electric kilns. Electric kilns facilitated individualized one-man operation (OMO) efforts compared with traditional timber-fired kilns that needed close cooperation between several individuals during the firing processes. The specialized skills involved in timber-fired pottery-making are described both by Moeran and my informants. While a detailed run-down is outside the scope of this chapter, it is possible to summarize the process in five stages. First, in the extraction stage for natural resources and ingredients, the potters' associates (suppliers in the case of the Hong Kong potters) extract clay, timber wood and other natural resources for making the pottery and for feeding the firing process. The second stage involves the creative process – designing and conceptualizing the forms and shapes of the ceramics. The third stage is the artistic phase where the ceramicists start to apply glazes onto the potteries and then leave them out to dry. The fourth stage involves the loading and firing stages where the potteries are loaded into the kiln chambers and then fired at an appropriate temperature. This may take place several times to achieve the desired glaze. The final stage is marketing and distribution where the final products are promoted to galleries, museums and shops and then distributed accordingly. In the case of the HK potters, stages two to five are managed almost exclusively by themselves as one-(wo)man-operation. Even stage one requires considerable efforts on their part to select the right suppliers and supplies.

While the potters' interviews with the author revealed the informants' opinions that their works represented solitary one-man-operation efforts rather than a specialized ecology, the author made three observations in his visits to their pottery studios that mitigated this argument. First, despite presenting the idea of autonomous work processes, the potters and their studios were located in fairly large artist clusters in industrial parks. The author visited two of them – the Chai Wan industrial park and Fo Tan industrial park facilities. The Fo Tan industrial park, according to Yim, hosts at least 100 artists, though another fourth-generation potter Ho claimed there were more. Wong revealed that he met the other fourth-generation potters at least once a month in gatherings organized in his stylishly-designed studio that came with a kitchenette. This indicated the presence of an informal ecology. Although not segmented according to highly-specialized roles, there was nevertheless a fair exchange of ideas, contacts and materials through such gatherings.

Secondly, there was substantial sharing of recycled products (please see the section on "Environmental Concerns and Practices") and gifting to each other. Some of these recycled products and gifts were directly connected with production work. The exchange of material goods is a tangible form of communitarian self-help. Yim and Lo also introduced some of his contacts to the project through this informal network. The most prominent example of mutual help that the author experienced in his fieldwork was in the creation of Yim's exhibition catalogue. A prominent Hong Kong photographer helped Yim create and arrange the visuals for this catalogue for only a token fee. The whole process took one year in order to capture the still objects on film under different seasonal lighting. The photographer's studio was proximate to Yim's workshop. The strong bonds between the two indicated that, not only was Hong Kong's artistic cluster closely-knit, but it had cross-disciplinary links as well.

Thirdly, the potters' ecological network is effectively globalized rather than local-community based like the Sarayama group. In her interview, Yim attempted to explain how Hong Kong potters effectively sourced and acquired materials and ideas from global sources and were not restricted by local boundaries. This appears to be a process of glocalization. On the one hand, some Hong Kong potters like Chen (Chan in Wades-Giles spelling) and Yu were discovering ways to indigenize artistic processes and natural materials extraction, but on the other hand, potters like Yim, Wong and Lo were busy trying to present the cosmopolitan side of Hong Kong potters' works and taking full advantage of artists-in-residency in places like the US and Japan. Some of the potters also travelled extensively to Burma, Taiwan, Tibet and Japan to get new inspirations and commissioned work.

In summary, I would argue that the Sarayama/Shigraki as well as Hong Kong potters are both located in ecological systems but with different characteristics. The Japanese empirical examples are based on specialized technical skills, a logical workflow from materials extraction to finished product and community-identity to prove one's worth to the collective. The resilience of the Japanese potters' ecological system is also based on time-tested traditions and clan ties in rural settings. On the other hand, the Hong Kong potter's ecological system is informal based on word of mouth and gatherings (the idea of *guanxi*), organically developed without specialized roles and highly diversified pool of skills. The Hong Kong ecological system also reaches across artistic communities to other technical expertise and enjoys global contacts across borders in sourcing for ideas and materials. Unlike the Japanese Sarayama/Shigarki community, it is urban in setting (in industrial parks and central business districts), compartmentalized spaces connected by word of mouth rather than close physical proximity of a village setting.

METHODOLOGY

In terms of methodology, this chapter relies on fieldwork, particularly oral interviews, for investigating the historical evolution of the contemporary art pottery community, understanding the production processes of the Hong Kong contemporary potters and comparing their observations of the Japanese pottery communities with their own. For the Japan component of the comparative study, I will rely on oral narratives of the Hong Kong informants' Shigaraki experience and secondary literature (such as Moeran's study) as well as observation studies in the summer of 2012 of Japan's pottery industry in general. In narrating my comparative work, I will minimize textual mention of Moeran's work except for the most crucial parts for comparisons due to prescribed word lengths for most journal submissions and readers can refer to Moeran's monograph for detailed ideas. The author also conducted fieldwork research in locations where pieces made by the fourth generation Hong Kong contemporary potters were displayed. For the visual portion of this chapter, the author also went to the Lamma Island and Stanley beaches where he was able to obtain surface finds as samples of utilitarian wares that were either part of trans-shipments to Hong Kong for re-export elsewhere or used by Hong Kong-based consumers locally. Their exact origins are unknown but may have been part of disposed thrash into the nearby seas or were discarded damaged wares from trading ships in the past.

I initiated my research project with library research and fieldwork. Funding was subsequently made possible through the South China Programme research grant. This grant made it possible to travel for fieldwork and hiring a research assistant. My research assistant and I conducted a comprehensive research of professional potters in Hong Kong and received some email responses to our online interview questions as well as permissions for face-to-face interviews. Through interviews, we were fortunate in having access to the ideas of Hong Kong's representative fourth-generation potters who are now moving into the most mature and prominent phases of their careers. I will profile our interviewees in greater details in the section below.

LITERATURE REVIEW

A reference used for this project was Brian Moeran's publication *Folk Art Potters of Japan* written about a local community of folk art potters of Japan. My interests were derived from a component of his project, in assessing whether potters in his study consciously thought about environmental issues as they went about making their crafts. This chapter's contents differ from Moeran's writing (a monograph) in three ways. I am adopting a comparative angle in looking at two groups of potters while Moeran's writing focuses on a specific local community in Japan (Sarayama). I intend to compare Brian Moeran's study of contemporary potters in Sarayama (Japan) with representatives of the fourth-generation contemporary ceramic artists in Hong Kong who studied and received cosmopolitan influences, including that of Japan (the Shigaraki pottery-making community). Moeran's Sarayama potters interested me because they were very different from my previous research conducted in 2010-2011 on the Imari and Arita export porcelain wares industry. Imari and Arita started off as small production centers like the Sarayama pottery community but expanded quickly to make up for global demand particularly in Europe in the seventeenth century. The Dutch introduced Imari porcelains to the Europeans to make up for the dip in blue-and-white ceramics supply because of internal political problems in China, including the drop in supply output from the then porcelain-making capital of the world, Jingdezhen.[6] In contrast, the Sarayama potters served mainly domestic audiences and experienced a golden period during the *mingei* boom of the late 1950s to the early 1970s[7] but they never reached an international audience in the same way as the Imaris and Aritas. They are small enough to be compared with the community of professional potters in the city of Hong Kong. By way of comparison, according to Michelle Chan, a reporter for *The Standard* newspaper who did an oral interview with the Chiu Kee porcelain factory (the last surviving family-owned pottery firm in Hong Kong), the 1960s and 1970s were also boom times for Hong Kong's indigenous porcelain industry.[8] Hong Kong pottery firms like Chiu Kee were able to expand to three workshops and hired dozens of workers to help out at all stages of the process.[9]

Unlike Moeran, I am not adopting an anthropological perspective to study an entire community multi-dimensionally. Instead, I am relying on oral materials in my case study of the Hong Kong potters to construct a historical narrative in a focused niche area. I am concentrating mainly on environmental and energy perspectives adopted by the potters in making their crafts. Moeran's publication (a monograph as opposed to this chapter) is far more comprehensive in painting a social portrait of the local community of potters.

HISTORICAL BACKGROUND OF CONTEMPORARY POTTERY MAKING IN HONG KONG

Hong Kong's history of pottery use and historical importation of export ceramics impact on the way the local contemporary potters conceptualize technologies, resource use, environmental impact and energy use.

Pottery Use in Hong Kong during the Shang Period

The history of pottery found in Hong Kong is datable back to the Shang period. Some of them have markings on the body that resemble those incised with a sharp object, cord or string. Shang period pottery in Hong Kong have similarities with those found in Japan in terms of technologies used for firing and also the use of cord or rope and incision for making patterns on the pottery bodies. If they were made locally, it meant that the local inhabitants already had primitive ideas and technologies about making pottery. If they were imported from the mainland, it confirms the notion that Hong Kong has had a long history of using imported wares even as early as the pre-dynastic times. From such early pottery use found in pre-dynastic small settlements, I will fast forward the story several thousand years later to late pre-modernity when Hong Kong grew and flourished into an important trading outpost in the late Qing period.

Pottery Use in Hong Kong during Late Pre-Modernity to Modern and Contemporary Periods

The author scoured the beaches of Lamma Island and Stanley area and used surface finds to date and uncover porcelain and pottery shards washed ashore or disposed in those areas which belonged to different historical periods. They are the remains of ceramics probably used in Hong Kong or were discarded from ships on route to trade elsewhere and many were made and imported from China rather than domestically-produced in Hong Kong. These imported porcelains are not the focus of this chapter but I am showing some samples here to indicate the existence of a history of imported ceramics for consumers in Hong Kong. I will also mention some implications of this historical trend below.

The Kitchen Qing Wares (Late Qing Dynasty Period) to First Half of Twentieth Century Found in Hong Kong

Large quantities of utilitarian wares imported into Hong Kong are at least traceable back to the late Qing period. The shards indicated in figure 5.1 are an example of kitchen Qing export wares. They are found all along the southern Chinese coast and throughout Southeast Asia, a major export destination of these utilitarian wares.

I acquired a kitchen Qing piece from an antique shop in Stanley to provide a visual sample of this genre of porcelains that were imported into Hong Kong in large quantities. The owner of the antique shop claimed that he sourced the bowl from a village in Guangdong. Regardless of the authenticity of the claim for its origin, it is representative of kitchen Qing wares used in southern China and throughout Southeast Asia. Chinese characters were included in the glaze designs in the center of the bowl on the body and at the bottom of the bowl. These characters probably refer to the following institutions: the studio that made these bowls; or the owners that commissioned them; or were intended as written texts for use in a ritual or rite, etc. Words were further incised by a sharp object onto the bowl, possibly to indicate the ownership records of this bowl, or intended for a ritual or other social functions.

The Postwar Period (1945 to Contemporary)

Not surprisingly, mid-twentieth century and postwar items were found in larger quantities amongst the surface finds compared with their late Qing counterparts. The tides at the beaches yielded large amounts of these broken pieces of porcelain. The collected shards dating from the twentieth century display several commonalities. They were mass-produced through a variety of techniques in terms of designs, including stenciled designs, printed pic-

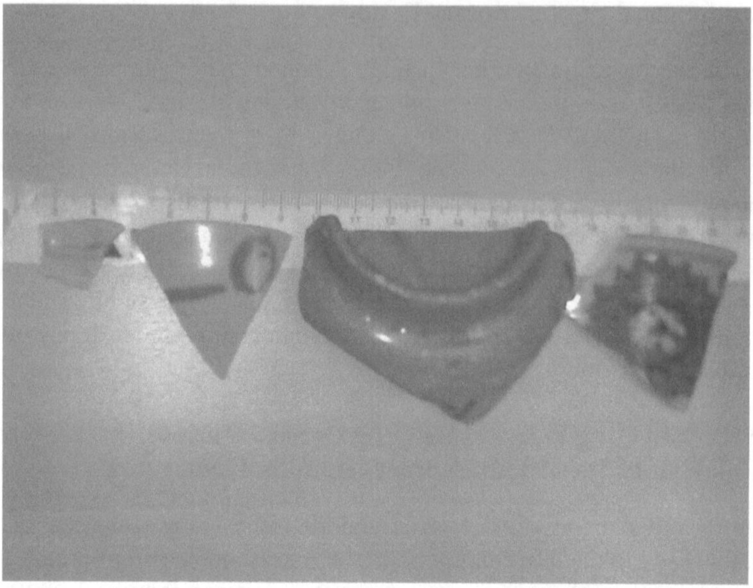

Figure 5.1. The shards in the photo are representative of typical kitchen Qing pieces. In the photo, the fourth shard shows a clear stylized design of chrysanthemum flower which was popular amongst export wares in that era. The rest of the pieces in the photo have typical amorphous designs found on utilitarian from that period. I picked up these shards as surface finds on the shores of Lamma island.

tures, hand-drawn lines, paper transfer designs and simply-applied glazes. It indicated a wide variety of wares for Hong Kong and other consumers to choose from. It also indicates the type of competition that any would-be potters in Hong Kong would face from products made and supplied from major existing ceramics suppliers.

The Twenty-First Century

To highlight the continuing tradition of ceramics importation in the twenty-first century, I acquired a few pieces of export wares from other countries outside China as samples. They are also indicative of the sort of high-quality competition that local contemporary potters had to face in promoting their unique utilitarian works and wares in this century. It is also indicative of the fact that competition in the Hong Kong utilitarian ware consumer market is now effectively globalized. Figure 5.1 indicates there is a long trading history in export ceramics (for example, the shards from the late Qing to contemporary periods). While importation of pottery into Hong Kong is established, this chapter is also concerned about contemporary ceramics domestically made in Hong Kong and the historical chronological account below is a survey of these artists. The contemporary HK potters that I studied in this chapter contrasts with those highlighted above in the following ways.

First, the large number of such pieces washed ashore and collectable through surface finds meant that there were large volumes of imported ceramics passing through Hong Kong in the past. Competition and also rising costs in Hong Kong have resulted in the disappearance of many large-scale pottery-making facilities. One of the few remaining ones which had a long history since 1928 is Yuet Tung China Works which is still operational and producing utilitarian wares. But the production site itself has shifted to China, probably to save costs.[10] The sheer volume of imported wares may have discouraged local ceramics production but it could also be the same motivating factor for local Hong Kong ceramics artists to focus on high-quality wares with value-added designs that are unique artworks by themselves. This in turn shapes the way contemporary Hong Kong ceramics makers have evolved into one-man-operations (OMO) with small-scale electric kilns that emphasize lean energy use and individualized care for resource use and work processes. Such OMOs are not possible in making mass-produced cost-effective utilitarian wares. They are the consequence of large-scale firings done in multi-chamber dragon or climbing kilns (for example, production of the late Qing to early twentieth century pieces). These large operations could also be carried out in large fossil fuel industrial-sized modern kilns (for example, manufacturing the twentieth century pieces) or large electric kilns (in making the late twentieth century and twenty-first century pieces).

Second, the richness and sheer variety of traditional designs of utilitarian wares from China may have prompted ceramics artists in Hong Kong to innovate their own indigenous designs. He Dajun, a pioneer pottery-maker detailed below, is an early example of how Hong Kong potters survive by developing innovative designs that reflected the cosmopolitan makeup of the city. He and other Hong Kong contemporary potters have infused East-West, Japanese influences with Lingnan culture to produce unique products. The quest for differentiation had an impact on resource use for some Hong Kong pottery-making pioneers. Indigenization prompted some Hong Kong ceramicists like Yu Dehua (Yu Tak Wah in Wades-Giles) to turn to local ingredients and resources for making pottery. These ingredients and resources were carefully chosen to produce unique pieces that are differentiated from their mass-produced counterparts. Yu even decided to extract raw materials for this purposes through his own means in Sai Kung. A more detailed discussion follows below.

The Pioneers: First and Second Generations of Professionals (1940s to 1960s)

Unlike Sarayama's rich culture in folk pottery, my HK informants tell me that there are few surviving folk potters in Hong Kong. They unanimously listed these folk potters by the handful. One of the first examples mentioned in the interview was He Dajun, a Shiwan pottery artist from the mainland who innovated on the orthodoxy of Shiwan traditions and produced outputs in traditional Lingnan Shiwan styles but with a Hong Kong interpretive twist. This was interesting to me, given that I was been mostly exposed to traditional industrial pottery from Shiwan (including roof tiles, architectural ornaments, flower pots and stone lions) and did not typically visualize such industrial ornamental pieces in artistic perspectives or as unique pieces of individualized artwork. I have also located undated Shiwan-style pottery in Sai Kung integrated structurally into Hong Kong's oldest traditional Taoist temple, Tianhou Gumiao in Hanyu Pinyin spelling (Tin Hau old temple in Wades-Giles). Shiwan-style tiles can also be found adorning the roof of a Stanley-area Tin Hau temple as well as its counterpart on Lamma Island.

From the perspectives of the fourth generation Hong Kong potters (my informants), there is a distinct limitation to traditional influences even amongst old masters like He who probably have the strongest links to the mainland in terms of stylistic influence. Hong Kong potters like He were willing to re-interpret historically strong traditions of industrial pottery in favor of individual styles that characterized the independent streak of Hong Kong's pioneers. This opened the door for subsequent generations of Hong Kong potters to pursue alternative options to preserving traditional orthodoxy. This contrasts with my informant Chris Lo's observation of the Shigaraki potters who stuck strongly to traditions or the presence of a Folk Craft Association which determined the authenticity of "old" shapes and designs of Sarayama (Onta) pottery in Moeran's study.[11]

> To classify whether the pottery in HK belongs to any type of 'folk' or not, I would say the 'Shiwan pottery' by the local potters are more 'Chinese' inherited then being very HK like. As for the rest of the HK pottery, since the impulse and spirit of works is a kind of mix of the East and the West and the local qualities, they look like 'folk-rock' or 'folk jazz' (pardon me for using musical terms)!

Here, Yim argues that Shiwan potteries are probably the most orthodox form of Lingnan traditional pottery adapted by earlier generations of contemporary potters. Local resistance to complete adherence to traditions further encouraged ideas of East-West fusions of pottery designs and styles in the works of later generations of potters. The routing of orthodox traditionalism in this statement create new spaces for Yim and her compatriots to continually innovate new ideas by adapting new technologies, develop new ideas about the environment and energy use. These points will be discussed later in this chapter.

Amongst the second generation of potters, unique characteristics of pottery-making individuals or groups began to emerge. While Hong Kong contemporary potters are mostly individual operations in general, there were some exceptions. For example, a family-based group of pioneer potters ran the Wunyao or Wan Yao in Hanyu Pinyin (literally translated the "bowl-making kiln") in Tai Po which has since become a preserved historical site. Their output classifies this family of potters as an industrial or utilitarian ceramics maker rather than purely folk or contemporary potters. The site produced mainly bowls, tiles and bricks. Another family group is busy protecting their traditional lifestyles, natural environment and handicrafts as the last family of porcelain painters in Hong Kong on Ping Chau Island. And within the family, only two of them are practicing the craft—China-born Lam Hon-chiu and Malaysian-born Lam Kew, collectively known as the "Chiu Kee" family enterprise. They are the only surviving potters on the island. According to the media, their survival was assisted by the Hong Kong Museum of Art who sought to preserve their crafts by persuading the Lams to organize ceramics painting workshops.[12] The integration of the Lams' lifestyles with their natural surrounding is visible in their enjoyment of afternoon hikes on the Island or catching their own fishes in the environment. This is a repeated theme which is visible in succeeding generations of potters like Yu Dehua mentioned in the section below.

Other than the emergence of family-based groups, amongst the pioneer potters, some technicians also stood out. A second-generation potter Chen Chunguang (Chan Chungong in Wades-Giles) was prominent in the 1950s and 1960s and owned his studio. Chen was well-known as an expert in constructing his own kiln. In fact, Chen helped my informant Yim with re-wiring and restoring one of her electric kiln:

> This kiln was constructed by Mr. Chan many years ago for one of his student. She [the student] passed it to me when she moved. Since its wire is quite worn out, Mr. Chan helped and taught me to re-wire and restore it about 2 years ago. Mr. Chan[13] used to be a very experienced kiln maker.

Chen was also a glaze expert who experimented frequently with glazes. Lo showed me a Chen-produced piece in his collection which was creatively glazed with jadeite stone powder, resulting in a mysterious, glassy and alluring glaze with a creamy greenish-white hue. The use of natural ingredients dug up from Hong Kong's indigenous lands indicate the convergence of indigenization of artistic concept and materials, materials found in nature and the emergence of a technically-competent generation of potters in Hong Kong.

Overall, from the chronology above, the first and second generations of contemporary ceramicists witnessed the emergence of technicians, family ecology and innovators whose output subsequently became valuable inspirations and intangible and tangible heritage for subsequent generations of potters.

The Western-trained Third Generation (1970s)

My informants agreed that Rosanna Li Huixian was representative of the third generation of Hong Kong potters who were influenced by the West and can truly be classified as the first line of contemporary potters. Educated in a Hong Kong Anglo-Chinese college, she proceeded to take up fine arts training in England. Her techniques came from the West although it is infused with some Chinese methodologies in pottery-making. Her concepts integrated the worldviews of Hong Kong people and the vicissitudes of Hong Kong people's lives. According to my informants, Li is probably the most notable public figure in their field as she often appeared in the media on topics related to pottery-making. Western ideas were not the only source of foreign influence for the Hong Kong potters. Influence from Japan also featured prominently in this area. Another representative of this generation is Yu Dehua (Yu Tak Wah in Wades Giles), whom my informants considered to be a mixed folk artist, who learnt pottery-making from his *sifu* (teacher), the second generation master Chen Chunguang. Chen imparted his skills to his best disciple Yu who subsequently moved to the rural district of Sai Kung in the countryside and set up his own kiln for making Japanese-inspired R*aku* ware.[14] Yu interpreted the *Raku*-style in a contemporary fashion and infused it with Western cultural influences. Again, his style infuses traditional Lingnan cultural influence.

The author has visited the Sai Kung countryside a number of times and found it an ideal environment for artists and lifestyle seekers with its pristine and clean environment that is relatively untouched by urbanization. According to the informants, Yu lives in Sai Kung in relative detachment and isolation. Perhaps, the most notable point about Yu is that he has a philosophy of self-reliance and dependence on natural resources. He dug his own pool in his Sai Kung residence, obtained natural ingredients for pottery-making from the ground and operated basically as a loner for more than 10 years. Yu is probably the closest OMO equivalent in Hong Kong to the collective Sarayama production system. Yu did receive support from members of the community, including first and foremost his landlord who admired his work and collected symbolic low rents from Yu as well as local residents who put up with his natural resource extraction activities. According to my informants,

Yu's work became an inspiration for others because it became an environmentally-friendly lifestyle choice and Yu's art-forms reflected his world-views on life itself. He was also admired for being self-reliant in terms of resource use and extraction. Yu was an inspiration for the third generation of HK potters onwards when there appears to be an indigenization process in pottery-making amongst the Hong Kong-based potters. Along with this indigenization process, there were traces of local ideas about resource use and environmental factors in their production and creative processes. My informants spoke about Mr. Yu's splendid isolation and indigenous extraction of resources with admiration and positive evaluation. This development seemed to be an early and distant source of inspiration for environmental consciousness and ideas about resource use amongst the succeeding generation of professional contemporary potters in Hong Kong.

The Fourth Generation of Hong Kong Contemporary Potters (Three Representatives from This Generation Are My Informants in This Chapter)

My informant Yim explained that the fourth generation of Hong Kong ceramists can be classified into two categories. The first group received indirect influence of *Mingei* Movement from cosmopolitan sources due to the international makeup of their teachers. For example, the graduates from HK Polytechnic- Higher Certificate for Studio Ceramic course, were taught under instructors from UK, the Philippines, US, etc. Some of these instructors themselves seemed to have absorbed the ideas and characteristics of the *Mingei* Movement masters. In addition to specific porcelain movements, the fourth generation of ceramicists also received influence from the contemporary art world at the same time (and this is not just confined to ceramic art field).

The second group of potters is the students taught by traditional senior *Shiwan* potters. And this group tries to integrate other influences with Shiwan traditional orthodoxy, resulting in exciting fusion styles which Yim suggests is a characteristic of cosmopolitan Hong Kong society. For this chapter, I interviewed Ms. Florence Yim, Mr. Chris Lo and Mr. Antonio Wong. All three are professional potters with additional daytime jobs for additional income. Ms. Yim and Mr. Lo teach classes on pottery to supplement their pottery-making activities while Mr. Wong is a marketing executive who also owns a pottery studio at Fo Tan. All three admitted that they could not make a living by simply being a potter but,[15] at the same time, they have had good opportunities to receive social legitimacy and recognition as their pieces were acquired by the Hong Kong Museum of Art, the tea ware museums in Hong Kong gardens located in Central and commercial hotels. This included the L'Hotel@Elan group who purchased Mr. Wong's pieces and the Hyatt Shatin which acquired Mr. Lo's artworks. Similarly, Ms. Yim has an impressive output record. Her works entered "The 8th International Ceramics Competition Mino, Japan" and are represented in the collections of the Hong Kong Museum of Art, the Hong Kong Heritage Museum, Shigaraki Ceramic Cultural Park, Japan and in private collections.[16]

The cosmopolitan influence of Mr. Lo's work is strong as they display innovative Westernized forms differentiated from basic Lingnan Chinese shapes and covered in creamy white glazes that feature firing bubbles integrated into the glaze. The glaze resembled the thick white Japanese Shino glazes or the glaze texture of Chinese *Jun* wares. The cluster of pottery on display at the restaurant is probably highly evaluated by the management of the award-winning restaurant who decided to showcase it in the restaurant glass display. The cosmopolitan influences found in Mr. Lo's works and his fellow compatriots are summarized as an observation by Yim:

> If they accept 'townsfolk,' also a kind of folk, they have to accept HK as a modern city has its own unique character quite different from the Lingnan heritage. Lot of local contemporary ceramicists, like many of us, are responding to the vibrancy and bombardment of this city through clay and formed our own 'contemporary folk' which is very diversified in styles — if 'folk' is meant to be non-static, but organic and able to grow and change alongside the time line.

This statement clearly implied the significance of continual innovation associated with the urban environment, indigenous characteristics of Hong Kong's cosmopolitanism that distinguishes itself from Lingnan orthodoxy. Yim admitted she and her compatriots' work are a response to the dynamics of the urban environment which is by itself characterized by diversity in expression of styles and influence. Yim appears excited by the prospects of un-ordered and slightly chaotic growth the urban environment in the city. This pro-innovation spirit and willingness to change old ways of doing things is perhaps the culmination of four generations of potters breaking free of Lingnan orthodoxy, embracing Western and Japanese influences in their work and initiating indigenization of styles by re-interpreting traditionalism. When the author and his researcher colleague Wong visited Yim's studio on 6 September 2012, he was able to see a panoramic view of a local neighborhood in the Chai Wan industrial estate.

To the author, the silhouette under cloudy skies of the meandering surrounding mountains enveloping a residential area in an industrial estate posed an interesting fusion of nature and the urban environment. But, Yim, who was more used to this environment having worked here for a period of time, highlighted the urban environment and imaginatively receded the mountain terrain into the distant background or cognitively de-privileged the mountainous terrain altogether. To the author, this was an example of Yim's ability to treat the aesthetics of nature and urbanity separately or together if she wished. The author intends to pursue such aesthetics and "sense" ideas about environment and nature in future writings. Similarly, the author intends to further pursue Yim and Wong's fascinations with natural forms such as butterflies through which they share a commonality. This common interest that is artistically interpreted differently by the two artists was sheer coincidence, according to Wong. Wong's recent works featured other natural forms like fishes, tail fins, trees and, most interestingly, the essence of Chinese classical conceptions of nature itself—Taoist balance between *yin* (female) and *yang* (male) forces in the natural universe.

Yim has also integrated yin-yang concepts in her works but conceptualizes it as a male and female virus entangled in a dialectic love-me, love-me-not concept.

Wong introduced fourth-generation potter Blanche Ho to the author and his research assistant colleague in his studio. Ho is the only fourth-generation potter to admit she has been substantially influenced by Japanese art trends and sense of aesthetics. Unlike Yu who makes Raku wares, Ho admitted she was influenced by Zen Buddhism-inspired ideas. Interestingly, Ho's understanding of Zen Buddhism is a Japanese interpretation but mediated through Taiwanese publications. For example, Ho is interested in the idea of simplicity represented through the aesthetics of minimalist clean lines of her pottery works based on the human belly which are integrated into furniture pieces. Ho's studio, which is also found in Fo Tan proximate to Wong's workshop, has a view of natural greenery which she enjoys in breaks between production work.

ENVIRONMENTAL CONCERNS AND PRACTICES

Their re-interpretation of traditional styles parallel ideas about environmental preservation and energy use. In terms of resource use, one difference between Japanese pottery communities and their Hong Kong contemporary ceramists is found in the production process itself. Florence Yim observed that, in Japan, the stages of production from extraction to technological management to creative production is a complete ecology by itself. Therefore, potters developed ideas of environmental sustainability (sustainable use of cedar trees for firing) in order to ensure constant supply availability of natural resources for production. In other words, the complete work process is constantly conscious in the minds of the Japanese potters. In contrast, the Hong Kong fourth-generation contemporary potters are knowledgeable about locations in Hong Kong where clay suitable for firing pottery can be found. But she and others within the community of potters in Hong Kong are unable to extract those clay resources because the small domestic demand for such clay does not justify extraction costs. Florence Yim commented that Hong Kong ceramicists are located in a "lonely" environment compared to the collaborative ecology of the Japanese pottery-making community.

Amongst the three Hong Kong potters, Chris Lo was the most vocal about environmental awareness. He was not keen on salt firing techniques because it produced acidic fumes. All three potters (Lo, Wong and Yim) had fumes and heat extractor systems that drew out heat and unhealthy fumes from the workshop out into the open space. Of the three fourth-generation potters that were interviewed, Wong is probably most inventive and hands-on when it came to designing contraptions for his studio. He self-designed a filtration system that can absorb the glaze spray particles during the production process through the filtration panels (wire-mesh item).

Toxicity

Lo also consciously avoided toxic materials including metallic oxide which would be poisonous if discharged into the sea. Likewise, Ms. Yim excludes lead glazes from her workshop for the same reason and relies only on non-toxic glazes. In sourcing purchases, Lo was meticulous in buying equipment and ingredients from worldwide sources selectively, clay from US, Australia and England, tools from Taiwan/Japan/China and machinery from Germany/Australia/US/China. He was careful to evaluate these materials and tools for quality first and foremost and other factors like toxicity. He was highly conscious of possible impact to the natural environment arising from their use, including items like chemical glazes, metallic oxides, types of fuel, etc. Lo also had to be conscious of the factor of toxicity because some of his wares are dual-use—displayed aesthetically but also usable together with other kitchen wares for meal consumption.

For example, the red terracotta pieces which Lo made are displayed aesthetically like pieces of a chessboard but can also be used as chopstick rests. They are made of non-toxic high-grade terracotta clay which can be infused with essential oil for an aroma therapeutic atmosphere at the dinner table. Lo also made sure that he only allowed his students to use non-toxic glazes. He is worried that, even if he highlights the toxicity of some glazes in class, students may forget about his advice after a certain period of time. Consequently, these students may then use the wares that they made in Lo's workshop for dining or gifting them to others for the same purpose.

Recycling

In terms of minimizing wastes, Chris Lo tries to recycle used materials as far as possible. In this area, Mr. Antonio Wong is in agreement with him as he argues that even the clay of unwanted pieces can be reused. Nothing is therefore wasted in the process. Lo's studio is well-organized in carefully categorizing the different types of unwanted clay in rattan basket trays lined with absorbent cloth so that they can be processed with the appropriate treatments. One mechanical way through which clay is recycled is using the pug mill which Lo had in his studio. The pug mill allows used unwanted moist clay to be regenerated into usable clay again. Lo who has the most number of students in regular classes compared with Yim and Wong has the greatest incentive to install this pug mill. By comparison, for example, Wong offers classes only on a project-basis and they usually integrate relaxation therapies workshops. Because Yim does not need to teach classes, she simply kneads unwanted moist clay on this plastic board so that she does not need to expend energy, finances and time on maintaining a pug mill. By comparison, Lo has large classes to teach and therefore, the rationale behind the acquisition and use of the mill. This is another example of how potters make rational decisions related to energy use in the production stage.

The biggest items that are recycled are probably the kilns themselves. Ms. Yim reveals this environmentally-friendly practice:

> The potters in HK have created a small community. Old or new kilns have been passed from one to another in order not to waste resource. (I have 2 other kilns; one given by Chris Lo and one by a school).

Yim is probably the most proactive potter of the three informants (Yim, Lo and Wong) in terms of using recycled kilns. She admitted having special affinities for unwanted old kilns. Yim has 3 electric kilns. One of them belonged to Ms. Yokky Wong who is Mr. Chan Chung Kong's student. Ms. Wong gave the kiln to Yim when the former moved to a new location. Another kiln came from Chris Lo and the third kiln received from a school. The school did not have a technician who could operate it, so the department simply wrote off the kiln from their assets and passed it on to Yim. An electric kiln in Yim's studio was gifted to her by a school that no longer needed it. At the point of transfer, it was 90% unused, in other words, an almost brand-new kiln for Yim. She has been re-using this kiln faithfully to fire smaller pieces. She makes this choice because it is more energy-efficient than some of her larger electric kilns. The other electric kiln in Yim's studio was also recycled and gifted to Yim by another fourth-generation potter. It was rewired by second-generation contemporary potter Mr. Chen Chunguang (Chan Chungong) for operational use. This act underscored the recycling and sharing efforts within the community of potters. Again, this is an informal form of communitarian ties that has some commonality with the much-tighter cooperative ecology in Moeran's study of Sarayama. The third recycled kiln in Yim's possession, is a gift from Chris Lo. This kiln is less energy-efficient than the newer models which Lo acquired for his current workshop. It is also less energy-efficient than Yim so she reserves it for firing larger pieces. Both Yim and Lo indicate conscious use of energy in the most efficient ways in producing their works.

Both Wong and Yim practice recycling unwanted wood pieces and by-products. Yim makes use of such pieces for her aesthetic work while Wong extensively uses used wood pieces for his studio interior design. As a cross-media artist, Yim shows the greatest flexibility in re-using natural materials. For example, natural feathers were used in her work to recreate the wings found in the Greek mythology *Flight of Icarus*. Reused materials are also used for production purposes by Wong. Unwanted pieces of wood are collected from a nearby carpenter's workshop and re-used as pottery-making platforms in Wong's studio. Wong also ordered a kitchenette set tailor-made from this carpenter so that fuel, energy and finances are conserved when the set is simply moved across the corridors. The kitchenette set was simply moved across a narrow corridor from the carpenter's workshop to Wong's studio. Planks made of chipped-wood by-products are used to frame the doorway leading to Wong's kilns. Wong purchased them as he appreciated the aesthetic value of chipped-wood pieces.

Simply Reusing

The simplest method of environmentally-friendly consumption of daily necessities and household or workshop items is probably using the same items for a long time. Lo's studio had some rags which he has used for a decade or so. Lo bought these mass-produced rags about a decade ago and still uses them faithfully, pro-longing the lives of these polyester non-degradable items as far as possible. The same goes for these moist rags which are stuff into the pug mill. They have been used for the same purpose as long as the machine's existence in the workshop. Like Wong, Yim also practices recycling. She collected rare *suanzhi* and *huali* wood pieces from cleaners who were on the way to dispose them. She also collected a Thai wooden rocking horse which was assembled with discarded pieces that used to be part of an ox-cart. Yim performs two roles here, recycling used pieces as well as saving handicraft pieces from destruction. Like his peers, Wong practices recycling by turning to the use of digital technologies, resorting to the internet to purchase used items. For example, the used sofa in the studio was acquired through the internet by Wong's friend. The bamboo blinds in Wong's studio are also secondhand acquisitions. The potter's stool in Wong's studio was a gift from Yim—a reused piece of production furniture that is itself made of raw unfinished natural wood pieces.

Education

Lo was explicit about long-term measures such as consciously imparting ideas of environmental protection to his students in his classes. In more philosophical terms, he urged the students not to produce low-quality pieces but to do serious work and make maximum efforts in every piece of artwork so that wastes in the form of discarded and defective pieces are minimized. Lo and others in the community even locate some of their classes in Hong Kong's environmentally-friendly spots like the Hong Kong Wetland Park which is a 61-hectare preserved area to offset the loss of natural lands that disappeared because of urban development in the Tin Shui Wai New Town to underscore the importance of the environment:[17]

> Ms. Fiona Wong, generation ceramist, has made used of clay in Tin Shui Wai, reformulate it and work out some potteries with children in educational function in HK Wetland Park (Chris Lo is also one of the instructors).

Other than purely production and educational issues, on a personal basis, environmentally-pristine locations were also valued by my informants. Yim felt that it would greatly enhance her work if she could be located in a place with beautiful natural scenery. This seemed to be a feeling shared by all three informants when they described Yu's Sai Kung kiln hangout. Lo recalled how the wide open spaces of Milwaukee Wisconsin inspired his works when he was an artist-in-residence doing observation studies in a Kohler ceramic factory. They were quick to add, however, that natural environments were a welcomed addition to urban settings in creating art products, and not completely replacing the urban environment. They reckoned that a city setting with its dense makeup and fast-paced rhythm is also a source of inspiration by itself for their creative process. However, at the same time, all three potters (Yim, Lo and Wong) softened the hard lines of the urban environment with indoor plant settings to add a touch of nature to the symmetrical lines of the city and their industrial workshops.

A view of the concrete jungle from Lo's workshop contoured with the soft lines of indoor green plants found in his studio. According to Lo, some of these plants also possess the quality of absorbing unwanted fumes and some radioactive elements emanating from the kilns. Lo's love for plants is evidently on display in the studio. He yearned to own a rural studio in locations like Sai Kung where he could have larger gardens to complement production work. The green contours and color of the plants softens Lo's studio view of a highly-urbanized concrete landscape characterized by highly angular and symmetrical lines. Inside the workshop, Lo places more plants and cut stems for the green touch. Nothing is wasted. Lo's cut lilies on display at the workshop counter. The cut leaves from these lilies are placed on the potter table for stenciling lines on unfired moist clay. This raw and primitive form of pottery design is traceable back to the Neolithic period when ropes, cords, leaves, knives and sharp objects were used to decorate clay objects before glazes were invented.

Aesthetically, Lo sometimes fuses natural plants with works of art. Pieces of fired pottery (gifts from a fellow potter) that resembles pebbles are placed in a strategic arrangement in this potted plant. The idea of gifting and sharing amongst the potters is a form of communitarian encouragement for each other's artistic pursuits. It is a looser form of cooperation compared with the close and specialized ecology of Sarayama pottery-making. In addition to aesthetics, Lo received information that the variety of variegated plant found in his studio has the qualities of absorbing unhealthy fumes and radioactivity emanating from the kiln. An aquarium, a microcosm of the natural aquatic environment, is on display in Lo's workshop. Aesthetically pleasing, it is a microcosm of crafted nature. From the author's perspective, Lo has been taking good care of the fishes, evident from the size of his neon tetra. Lo admitted the tetra fishes are much larger than the average size for this species due to their age under his care.

Recycled Décor

All three potters have a penchant for using recycled items to decorate their studios. The ideas of recycling appear to be integrated with the three potters' lifestyle choices, sense of aesthetics and also commercial displays in the workspace. Other than organic wood pieces, recycled decor also included plastic items. The plastic heart and red-star decorative items decorating Wong's storeroom and office space were collected from a thrash depot. Sometimes, organic wood items and recycled materials are integrated into the art form itself. Besides aesthetic displays, Yim's mixed media use meant that she had a number of natural objects found in her workshop for design inspiration. The branches and stems of a dried plant hold pottery pieces together in Yim's multimedia artworks. The natural forms of branches, stems and twigs found in her workshop are also referenced for her artistic works. Natural wood pulp fibers are used to create Yim's intricate tissue-paper ceramic design. Chinese medicinal herbs are also integrated into this art piece and the herbs give way naturally in the process of firing, leaving the core ceramic structure intact. In her current project at the time of the interview, Yim intended to paint long-pea pod casings that she collected at The Chinese University of Hong Kong (CUHK) campus for creating a multimedia work. The author found similar dried long-pea pod casing in Wong's studio although he uses it mainly for aesthetic display. Wong collected them at Victoria Park in Hong Kong. Yim also collected coconut husks and dried gourd from natural environments.

Energy Use

The Hong Kong potters' experiences in Shigaraki and Moeran's Sarayama community point to three important features of Japanese pottery-making practices. First, productivity and efficiency were not the only consideration in making decisions about fuel use. There were strong desires to continue traditional practices even if they tended to be less productive and sometimes energy-efficient. Moeran mentions his feature in his Sarayama study: "However, so long as they kept using the old-style sand-sloped kilns, there was a limit to the number of pots that could be made and fired at any one time."[18]

This observation was confirmed by my interviewees. From his study of Shigaraki, Chris Lo noted that wood-firing took at least five days to heat up and cool down. The *anagama* kiln in Shigaraki, he observed, was really only used two times a year and the rest of the time was spent on extracting timber from the mountains. Such lengthy time period spent on preparing the *anagama* for firing, in his view, were probably not tolerated in fast-paced Hong Kong. Therefore, in Lo's view, this process was not efficient, economical and energy-saving. Moreover, Lo felt that wood-firing was not environmentally-friendly in itself due to carbon emissions. When he owned a traditional timber-fired kiln in Ping Chau Island, Lo was careful to use recycled wood pieces and timber for his production activities. However, even though Lo was environmentally-conscious in this aspect, he realized that recycled wood from used furniture found in Hong Kong and the mainland contained too much varnishes and toxic binding glue which released toxic and excessive smoky white fumes when burnt as fuel in the kiln. This was a factor that prompted Lo to give up his traditional kiln in favor of electric kiln use. Wong also revealed that Hong Kong regulations did not permit timber-fired kilns to be operational due to the same reason of gas and smoke emissions.

From a non-technical perspective and one that focuses on the spiritual dimension of timber firing, Wong talks about his own experience of working in Shigaraki. The morning scenery with its chilly misty appearance and dew drops as well as the sunrise was a natural inspiration for Wong. Wong felt spiritually enriched chopping his own timber pieces in preparing for kiln-firing. He had a special feeling handling natural ingredients with his own hands, something not experienced in electric kiln-firing. Through this process, he felt proximate to the natural environment. It is not known if Lo's and Wong's views are commonly-held amongst Hong Kong's potters in general but Yim did reveal that there is a tendency for Hong Kong potters to go overseas to nearby locations when they intend to use wood-fired kiln technologies for their works:

> A lot of Hong Kong potters has been trying to overcome the limitation of resource and support or expand the choices of resource by looking for wood-firing kiln sites in Mainland China like Shiwan, Jingdezhen, 龍泉 (Longquan),

宜興 (Yixing)....etc, those of Taiwan, Japan (for example, Shigaraki, Tokoname). Some also went to Australia, etc to do firing. …Another second generation potter, Ms. Wong May Lee (non-English speaker) is the director of "Tao Tao Fong" pottery workshop, she has been organizing a number of wood-firing tours to mainland China and Taiwan. Mr. Ho Tai Kwan has also been doing similarly."

Yim showed me an example of wood-fired burnt marks on a ceramics pieces in her workshop produced in Shigaraki. The technically-created burnt marks are only achievable in timber-fired kilns.

An interesting question that I have with regards to foreign-location firing is if the process would yield new creative results. In other words, would Hong Kong potters who use their own techniques or ingredients to fire in mainland Chinese or Taiwanese kilns produced new innovative pottery? Especially since the energy produced by the flames of a timber-fired kiln are likely to be different from those of an electric kiln favored by my fourth generation potter informants. Again, I intend to pursue this in a future paper or chapter on the aesthetic implications of energy use.

A major factor that shaped the fourth-generation Hong Kong potter's energy use and choice of electric kilns is the desire to lower carbon footprints and emissions. But Lo opined that the Japanese potters continue to preserve traditions even if newer designs made it possible for the potters to have better efficiency. He did, however, admitted that the cedar/pine trees that the Japanese potters used as fuel for the kiln firing had a soft effect and produced longer flames. These features were important in creating certain interesting effects, a point elaborated by Yim: "The longer flames in wood-firing produces interesting "smoke and fire marks" which give unpredictable shades and textural quality on clay body of the pots."

In other words, there are technical considerations to energy use rather than merely cost factors alone. But in my conversations with my informants, the cost factor remains highly important. For example, for Lo, his main consideration behind energy use is the cost factor. He argued that he was careful in tracking energy consumption, especially its consumption by the electric kilns and air-conditioners found in the studio that he owned. They were the two main sources of energy use and costs incurred in the use of electricity. He was conscious of shutting down both appliances when they are not in use to conserve energy. Another energy-saving practice is the avoidance of chamber-firing until the kiln is fully-packed with unfired clay. Wong does the same to save energy costs even though he is a solitary artist. Both Lo and Wong are conscious about the high-intensity energy use of electric kilns. Amongst the three fourth-generation potters that I interviewed on 31 August 2012, Lo was probably the most energy-costs conscious due to his work in teaching full-time classes (10 students per session) compared with the other two fourth-generation potters (Wong and Yim). Therefore, energy bills for Lo are naturally higher than the others.

Lo never spends time to pre-heat a kiln for wet pots to conserve energy. Lo does not start firing until all pieces are naturally dried and does not pre-heat the pieces. The precision in which these firings take place is only possible because of accurate temperature controls made possible by the electric kilns. Both Lo and Wong use the same American-made kilns (Lo also owned German-made ones) which they claimed is energy-saving. Lo revealed that newer models of the electric kilns are remote-controllable through downloaded applications in the smart-phones. The applications can manage the kilns' temperature through alerts received via the smartphone system. These kilns were easy to use and more cost-effective compared with building their own kiln (which Yim had attempted). In their view, there is greater control of the firing process.

CONCLUDING SECTION

From the two localized studies of important pottery-making communities in this chapter, I have confirmed the importance of environmental and energy factors in their production processes. In many ways, the deliberations made by the HK potters related to these two factors appear to be utilitarian in nature—that of saving costs and ensuring the sustainability and stability of supply of raw materials and ingredients for the firing and potting processes. However, in the case of the Japanese pottery communities in Sarayama and Shigaraki, cost-effectiveness is not the unmitigated dominant factor influencing the production process. Traditional cultures and ways of doing things continue to resist complete conversion to and reliance on modern technologies and methods of production. However, the important caveat in this area is that my study does not cover industrial pottery-making or large-scale mass manufacturing which are likely to turn up different observations and conclusions.

NOTES

1. This chapter is funded by the South China Programme Research Grant.
2. Moeran, *Folk Art Potters of Japan*, 2.
3. Ibid., 1.
4. Ibid., 53, 103.
5. Ibid., 207.
6. For updated accounts of this story, I recommend referring to Shimura, *The Story of Imari*.
7. Moeran, *Folk Art Potters of Japan*, 140.
8. Michelle Chan, "Last of the Porcelain Painters," *Standard*, January 20, 2007, accessed September 5, 2012, http://www.thestandard.com.hk/news_detail.asp?pp_cat=30&art_id=36430&sid=11773593&con_type=3.

9. Ibid.

10. Yuet Tung China Works, "About Us," accessed September 5, 2012, http://www.porcelainware.com.hk/aboutus.html.

11. Moeran, *Folk Art Potters of Japan*, 169.

12. Chan, "Last of the Porcelain Painters."

13. According to my informants, Mr. Chan has since relocated to the US.

14. Yu is not the only Raku-influenced potter. Chan Kam Shing and Yu Tak Wah have also constructed their own *raku* kilns.

15. It was not always this way in Hong Kong. In the late 1950s, a construction laborer made HK$3 daily but a porcelain painter gets up to HK$200 monthly. Foreign competition and rising labor costs are likely factors for diminishing the pays of professional potters. Michelle Chan suggested specifically that competition from post-reform China ended Hong Kong's pottery industry. See Chan, "Last of the Porcelain Painters."

16. Wai Wai Yim, "Yim Wai-wai – [Art] Attack 16," last modified August 18, 2009, accessed September 5, 2012, http://www.choochootv.com/choochootv-shows-/artattack/195-yim-wai-wai-artattack-16.

17. Hong Kong Wetland Park, "About Us – Background," 2012, accessed September 5, 2012, http://www.wetlandpark.com/en/aboutus/index.asp.

18. Moeran, *Folk Art Potters of Japan*, 126.

Chapter Six

Post-Modernism and Pottery-Making

A Case Study of Issues of Artistic Production, Environmental Awareness, and Energy Use in Hong Kong with Comparative References to China and Japan[1]

In the previous chapters, the progression of fuels transitioned relatively unproblematically from wood fuels to more combustible and efficient fossil fuels. This deterministic progression is mitigated by other factors for consideration such as environmental concerns and post-modern ideas about consumption. Based on an oral interview and a study of related narratives, this chapter is interested in discussions of post-modernism in Hong Kong with comparative references to China and Japan. The literature review looks at some narratives in existing literature related to the idea of post-modernism argued by scholars in the context of Japan and China. There are three aspects of ceramics production for focused discussion in the chapter. First, how is post-modernism defined in Hong Kong by its ceramicists and how is it different from mainland China? Second, based on the oral interviews with Hong Kong-based contemporary potters, how does artistic production of ceramics in individual studios compare with mass-produced pieces? Third, are energy and environmental concerns related to post-modernism and are they conscious factors for consideration by the artists involved in the production process? The methodologies used in this chapter are based on textual analysis and oral interviews with contemporary Hong Kong potters. The chapter will begin with a literature review of the idea of post-modernism based on comparative studies of China and Hong Kong before moving on to analyses of oral interviews conducted with Hong Kong-based potter Ms. B and her comparative experiences in China and Japan.

LITERATURE REVIEW

In the case of China, post-modernist impulses are often studied through the empirical example of Hong Kong because of its status as a cosmopolitan city that has enjoyed a long period of postwar economic development as a market economy. It is characterized as having an advanced stage of capitalistic development as one of the four dragon economies. Arif Dirlik and Xudong Zhang edited an important volume that examines postmodernism in China and Hong Kong. In their introduction, they looked at the relationship between post-socialism and the rise of materialism, consumption and the market economy for China. However, the volume differentiates Hong Kong-style consumption and postmodernism from post-socialist Chinese ideas about postmodernism. Dirlik and Zhang's edited volume positions market-economy Hong Kong as a component of China unrelated to experiences in socialism or post-socialism. Hong Kong's postmodernist experience is closely associated with capitalist consumption prioritized over all other ethical concerns. Perhaps, one of the most extreme interpretations of Hong Kong's capitalistic tendencies comes from Xiaoying Wang: "To a population that was already predisposed to the activity of profit-making, there was no need for a systematic ethical neutralization of the profit motive to remove moral barriers to legally permissible profit making."[2]

The accent in Wang's narrative is the dominant influence of market forces in crafting Hong Kong's society and its values. This statement suggests the predominance of capitalistic instincts in Hong Kong to the exclusion of ethics, morality and other non-profit concerns. Viewed from the left or socialist perspectives, this critique of Hong Kong postmodernism portrays a relentless capitalist economy that has lost its sense of ethics and morality, holding closely to material progress and material-based quantitative improvements. This perception of post-modern Hong Kong is a popular narrative in popular and trade publications, media and street-level discussions related to its laissez-faire economics. To test its applicability in different contexts, I decided to isolate one particular sector in Hong Kong, the pottery-making community, for further discussion in this area. My fieldwork with Hong Kong contemporary potters appears to indicate a complex interplay of pragmatic capitalist concerns (making a living, profitability, sustainability of traditional crafts and micro-entrepreneurship) with the informant's post-modernist ideas (self-actualization of artistic goals, hand-made products, and unique artistic pieces that are not mass-produced). My informant's strong accent on pragmatism and non-material pursuits is, therefore, a mitigation of extreme readings of Hong Kong either as a merciless capitalist society or an advanced economy that has transcended economic pursuits. The reality is an eclectic combination of both priorities as potters craft their unique wares in relative detachment from market demand and profit-making expectations while coping with practical livelihood issues like rent and cost of living. I will discuss these complexities in the oral interview section.

From the pro-market, pro-capitalism end of the political spectrum, this interpretation is a natural feature of Hong Kong's laissez faire economy whose prosperity depends on continued absence of state interference in its business sector. It sees nothing fundamentally wrong in maximizing profit-seeking. In reality, however, post-modernist ideas in Hong Kong is likely to be located somewhere in between these two ends of the ideological spectrum, often an outcome and compromise reached in the contestation of ideas in the intellectual, social and policy-making spheres. In the natural course of its capitalistic development, social concerns about income gap, housing availability, environmental preservation and egalitarianism are likely to resist complete non-state interference and non-moralistic profit-seeking impulses. These social concerns are likely to become natural limitations to unbridled capitalism. On the other hand, post-modern attempts to resist capitalism are likely to be countered by a more compassionate, consultative, consensual and participative form of capitalism. Capitalism in Hong Kong and its record of success are likely to evolve new features to co-opt post-modernist impulses and neutralize its confrontational elements. A more compassionate form of capitalism may include equitable distribution of financial handouts to residents, policy provision of more land for housing projects, augmenting social safety net features, growing attention to corporate social responsibility (CSR) and "green" ways of doing business, etc. Expressions of these ethical concerns are much more complex than it appears. Instead of having clear positions, I am inclined to describe Hong Kong post-modernism as a form of indigenous pragmatism moderating extreme interpretations rather than unambiguously taking a pro-capitalist or anti-materialism stance. Figure 6.1 shows a visual representation of this delicate balance. I will discuss how this balance is managed, navigated and crafted by the contemporary Hong Kong pottery-making community in the subsequent sections.

HOW DO THE POST-MODERNIST NARRATIVES AFFECT CRAFTS-MAKING?

Narratives that support or are critical of postmodernism have three implications for the ceramics industry. First, consumption and production go hand in hand. Modern mass production systems originated from the ideas of Fordism and Taylorism which systematized manufacturing techniques and standard-

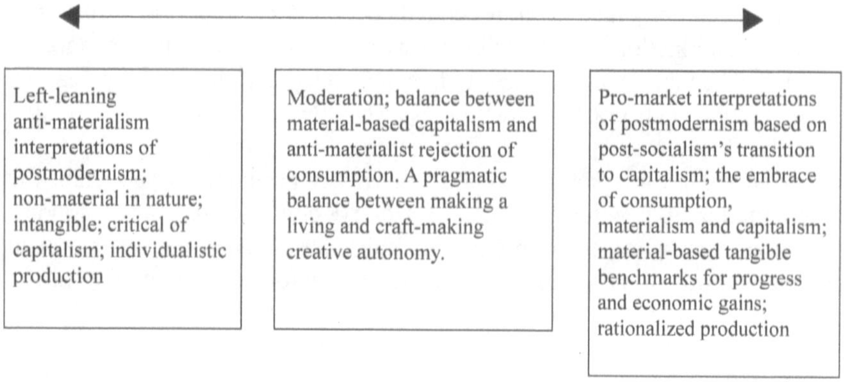

Figure 6.1. Author's own diagrammatic spectrum of interpretations of postmodernism.

ized work processes to maximize output. Mentioned in the historical overview of this chapter, the deterministic evolution of fuel use towards environmentally-friendlier, greener as well as greater control of the firing process in high-tech kilns is necessary to produce high-quality standardized utilitarian wares with precise dimensions and even glazes. Timber-fired kilns or smaller folk kilns are unable to achieve this large-scale production with comparable quality assurance. Similarly, mass-manufacturing techniques and mass-produced kitchen wares are representative of household products which are consumed in large quantities, symbolic of an advanced capitalist society and economic progress.

For my interview work, I wanted to understand ideas about post-modernism from a Hong Kong-centered perspective and specifically how individuals in the territory understood it subjectively. I chose to focus on an oral interview case study of a fourth generation Hong Kong-based contemporary potter, Ms. B. I carried out this oral interview on October 18, 2012. By the time that I interviewed Ms. B, I had already carried out interviews with four of her other peers in Hong Kong. Ms. B's contact was obtained through my Research Assistant (RA) colleague. Ms. B in turn then introduced Ms. A and other prominent fourth generation contemporary Hong Kong potters who also have rich experiences in the territory's vibrant art scene for soliciting oral interviews for the project. They represent a new generation of potters who are coming into prominence in the Hong Kong art scene now with works featured in gallery shows, museum collections and the media. Ms. B's interview is unique and stands out for three reasons. First, she was the only contemporary potter in Hong Kong that I interviewed who had distinct personal ideas about post-modernism. Second, she had strong consciousness of the environmental impact of ceramics-firing in Hong Kong, Japan and China. Third, she had rich ideas about the concept of "folk art" from her own understanding.

Ms. B's subjective ideas about post-modernism can be divided into three components. The first component emphasizes her opinion that post-modernism is contextual, situational and fluid, definitions of post-modernist ideas is based on how an individual applies it to interpret or analyze an idea. Because of post-modernism's amorphous nature, Ms. B feels that the concept is too difficult to understand, excessively abstract and confusingly odd compared to modernism. In her view, the concept of modernity by comparison is tangible. Modernization produced goods and objects that one can touch and visualize physically. In this sense, Ms. B opined that post-modernism is not objectifiable but based on individual ideas and idiosyncratic subjective cognitive awareness. By clearly defining modernity and marginalizing it from what is not modern, Ms. B's idea of post-modernism is derived by differentiation from modernity. The latter is tangible and quantifiable but the former is conceptualized as something intangible, amorphous, non-physical and based on cognitive awareness rather than visual experience.

Ms. B tries to characterize post-modernism in her works as the subjective understanding and application of traditional values and culture. She defines this through comparative means. Because of her artist-in-residence in Japan (specifically Shigaraki ceramics park for three months), Ms. B developed a comparative perspective of traditional Chinese and Japanese cultures. Through distinguishing her pre-existing notions of traditional art and culture with new ideas that she found in Japan, Ms. B was able to craft her own interpretation and understanding of what she regarded as "traditional culture." According to her, the idea of "traditional culture" comes in various forms. It could be intangible like the virtues and moral values expressed through art-forms like calligraphy (another one of Ms. B's specialty) and pottery-making. These values are defined broadly in terms of what makes a person good and virtuous and it was dependent on Ms. B's subjective and situational readings. She was unable to provide distinct benchmarks, barometers or determinants for judging morality or virtuous behavior. The morality that Ms. B referred to appears to be amorphous and based on her own contextual interpretation of what is considered positive in life and in the universe.

Ms. B also discussed the idea of traditional cultural values in aesthetics. She uses a comparison to make this point. In her view, the aesthetic difference between a Japanese and Chinese gourd lie in the extent of "fullness" of the vegetable shape. According to this interpretation, the Chinese gourd emphasizes "fullness" while the Japanese counterpart has a narrow construction and a narrow waist. The latter, considered as foreign and alien (what Ms. B terms as "weird") is marginalized here, breaking off from Ms. B's conventional understanding of the "traditional" shape of a gourd existing in reality as well as depicted in ceramic works. The consciousness of the differential construction of gourd shapes was reflective of her understanding of "traditional" versus "new" ideas found in her works and others. Ms. B's gourd analogy differentiating between traditional and new ideas paralleled her cognitive boundaries between modernity and post-modernism. Both were marked by exclusionism – the gourd object without a full body was excluded from her aesthetic conceptualization of a traditional Chinese gourd. In the same way, she excluded what was not observable, empirical and tangible from modernity and by exclusion, whatever fell outside these categories were defined as "post-modern." Her reading of post-modernism does not seek social consensus or doctrinal verification from others within her community. It was subjectively based on her own cognitive categorization through exclusion.

Besides ideas about aesthetics, traditional values are also associated with spirituality in my oral interviews with Ms. B. Ms. B is an atheist who was not affiliated to any religion but installed "kiln gods" on her electric kilns for luck. I encountered similar kiln gods in my fieldwork in the summer of 2011 installed on an *anagama* in Izumi city. Here the Izumi *anagama* owner attributed the presence of kiln gods to Shinto traditions, performing rites with sake offerings made to the gods to pray for successful firings. Ms. B's cited ideas about spirituality and her reasons for displaying kiln gods in her studio were different from the *anagama* kiln that I encountered in my Izumi fieldwork. It was non-religious, had no creationist features, or the concept of a higher being. Rather, the kiln gods were beings with no tangible form or existence but one that guaranteed greater successful kiln firings without any religious explanations or the need to perform any rites.

In fact, the kiln gods did not even have a well-defined explanation of how they attained deity status in Ms. B's beliefs. One of the figurines was a kiln god gifted from an acquaintance in Jingdezhen while the other two were unconsciously created through accidental firing. They were not made on purpose. These "gods" came into being without conscious design but nevertheless served as emotional support for Ms. B's production. While this form of spirituality was not directly related to her works in the production or aesthetic sense, it was a form of cognitive rationalization, emotional dependence and psychological relief that Ms. B relied on when firing her works. The figurines were simply there because it was appropriate to put them there. Just like the Chinese gourd was automatically attributed a full shape, there are no rational explanations. They were defined subjectively according to cognitive awareness interpreted through direct visual contact and not intermediated textually.

The second component in Ms. B's idea of post-modernism is related to aesthetics. Like the other fourth generation of contemporary Hong Kong potters, Ms. B distinguishes her works from mass-produced products. Ms. B and her peers' crafts emphasize individual design by incorporating unique ingredients and materials not found in standardized mass production. Collaborative works rely on informal community-based cooperation rather than a massive industrial superstructure that can quantitatively churn out standardized goods. Pottery production by the fourth generation of Hong Kong contemporary potters are artistic responses to mass production and consumption. It is also not rationalized like mass production because the emphasis in such works is not productivity and efficiency but individual expression of artforms.

Ms. B provides her interpretation of the historical background behind the origins of crafts production in Hong Kong. Historically, in Ms. B's view, the origins of folk pottery in Hong Kong were pioneered by Ms. Mak Yee Fan who developed her career in UK. To learn more about Mak and to corroborate Ms. B's oral materials, I accessed her official website. In her official website, two features on the website seem to be related to post-modernism. First, Mak highlighted her preference for simplicity in production matters,[3] as opposed to the presence of elaborate divisions of labor and high levels of mechanization found in mass production. Second, Mak's reductionist interpretation of reality as a single dew of water carrying the image of the entire world[4] corresponds with her preference for simplicity in production methodologies.

Mak's ideas represent a conceptual rejection of complex built-up systems of production. The underlying philosophy that Mak adhered to forms the conceptual foundation for her works. The idea of a concept underpinning a series of works is an important criteria for Ms. B, the informant in this chapter, to distinguish between crafts and household wares. Ms. B also argues that art needs to be created for a purpose. In her view, a design concept needs to serve functionality, conform to a philosophy and not necessarily serve only aesthetic pleasure. Before Ms. Mak's works appeared in the Hong Kong art scene, according to Ms. B, folk art was mainly produced by amateur potters in the form of utensils, daily wares or other items for daily or commercial use. Handicrafts have the potential to reach folk art if their design and form conformed to a concept.

If Mak was a pioneer in creating folk art in Hong Kong, then in Ms. B's view, Lucie Rie popularized it. Lucie Rie, an internationally well-known English artist of Polish origins, integrated folk art designs into household utilitarian wares and introduced these works to the Hong Kong fine arts community in the 1980s. In terms of international recognition, Rie's artworks were legitimized through acquisitions by international museums. Corroborating secondary resources and online information, I did research on Rie's biographical information and found that it matched Ms. B's information given during the interview. Rie was featured widely in trade publications and the international media like *The Guardian*[5] as an international icon for transforming seemingly ordinary household items into avant garde conceptually-complex design art pieces. Ms. B captures the essence of Rie's pioneering work in arguing that, without a design concept behind a piece of work, an item like utensils cannot be considered as crafts or art. Another person of influence on the development of folk pottery designs in Hong Kong cited by Ms. B was Katherine Mahoney, an Australian who specializes in handmade pottery pieces with her own official website at http://katherinemahoney.id.au/. The section of her website on ceramics provides details on how she started making art works. From my visual examination of the pieces that she displayed on the official website, I spotted functionality, simple lines, a pleasant feel, pastel colors and ample use of brushwork splashes and dripping techniques to enhance the aesthetics of the utilitarian wares. Ms. B mentioned that she and other artists in Hong Kong received influence from Mahoney's interpretations of design forms and colors of functional wares.

According to Ms. B, however, kitchenware crafts have waned in prominence since the time of Mak, Rie and Mahoney. But both crafts and household utensils are subject to the same powerful force of fickle-minded fashion trends. Ms. B attributed the decline of folk art interest in kitchen wares designs in Hong Kong to consumers and art patrons who had moved on to other types of art works to satisfy their consumption and collecting impulses. Here the realities of capitalist consumption are real and continue to define the boundaries of artistic and aesthetic creations. Besides the impact of fashion trends, crafts-making runs into another formidable rival in the form of mass production discussed below. Despite the decline of folk arts in Hong Kong relative to its heyday during Rie's time, folk art utilitarian wares pieces opened the way for individually created pottery works of arts for consumption by the artistic, and collectors' communities.

CRITIQUE OF MASS PRODUCTION

The third component of Ms. B's ideas about post-modernism is based on her ideas of mass production. Here she engages in discussions about capitalism. Ms. B tackles the idea of the tendency for mass production to generate surplus through excessive production. Contrasting handicrafts with utilitarian functional wares, Ms. B lamented that there are too many mass-produced functional wares in the market today. Her philosophy is centered upon moderate consumption based on actual needs. She did not believe in inventory stock-up or surplus production and consumption. In her view, only a small minority of mass-produced wares have good quality.

In her private circle of friends, Ms. B makes it a point to use and share mostly hand-thrown pieces only. Even in her social interactions with others, Ms. B practice what she preaches, for example, gifting her hand-made pieces to her mates for weddings and not relying on conventionally packaged wedding gifts manufactured for the masses. In emphasizing the value of handmade pieces, Ms. B invokes the element of nostalgia. Ms. B remembered the earlier days in her life when mass-produced items were not as prevalent and she treasured the things that she owned and acquired. She believed in the need to preserve and maintain personal artifacts in good condition to keep them in use longer. Ms. B mentioned that China in the past as well as in contemporary Japan, broken pieces would be mended by some craftsmen and they would still remain in use thereafter.

Ms. B relied on aesthetic concepts to differentiate mass production from hand-crafted creative products. Overall, in Ms. B's view, the core differentiation between art and crafts versus functional kitchen wares is the idea of a concept. If a particular ware or its series have a conceptual foundation in the aesthetic and functional sense, it can then be considered as crafts. By comparison, she argues that household kitchen wares merely have colors and forms but not an artistic or design concept. But she warned that, like commercial items, both crafts and functional household products are equally susceptible to fashion trends. Once the trends are over, the popularity of the products then recede and go into obscurity. The close association of the craft with commercial concerns and capitalism appears to be an omnipresent force in Ms. B's world. Even though she rejected the value of commercially mass-produced goods and questions the idea of abundance or excess in overconsumption, she acknowledged that it was a challenge to sell her works as the market for ceramic pieces in Hong Kong is limited and the pieces consequently are difficult to market and sell. The only real avenues for sales, albeit limited in scope, are exhibitions and galleries. Ms. B did not do any promotional and marketing efforts in her practice. Like the other ceramic artists that I interviewed, she mentioned that paintings, on the other hand, were comparatively easier to find consumers.

MS. B'S POST-MODERNISM: TRANSCENDING CAPITALISM

To transcend the profit-making aspect of mass production and the prevalence of standardized designs for the masses, Ms. B needed to transcend the idea of capitalism itself and come up with new ideas to explain her motivations in pottery-making. Here, interesting ideas related to post-modernism arises. Like the discussion in the literature review, Hong Kong does not exhibit the kind of post-socialist transitions that China faces. Instead, Hong Kong's advanced stage of capitalism centers social discourses around the ideas of consumption, commercialization and their social impact. In the wake of China's economic reforms after 1979, the desire for wealth acquisition has given consciousness to the idea of class and income gaps in post-reform China.

But, because Hong Kong is ahead of the mainland in her current stage of economic development, Ms. B's thinking about postmodernism indicates she is not class conscious. Her only concern is to assuage anxieties to meet her basic financial needs so that she can concentrate on her craft without worrying too much about daily expenses.

She self-rationalizes and neutralizes any worries about difficulties and challenges in livelihood related to pottery-making by referring to the activity as a higher form of self-actualization, that of making art for its own sake without overemphasis on its profit-making value. In Ms. B's view, art is not equated with profit orientation. She was willing to give up a full-time job which would have brought her more income. Indulgence in this higher form of needs, Ms. B admits that she required financial support from her family members. This was a common point mentioned by other contemporary Hong Kong potters that I interviewed. An alternative route that some artists who had no family support rely on was to conduct pottery classes for fee-paying students. Like the discussion in the literature review, Ms. B's self-rationalization skipped any internalized debates or objections about the necessity of profit-making and instead focused on non-profit considerations, that is, meeting her own interest-driven objectives and the aesthetic values of her work. The limited market size and profit-making opportunities indicated by all Hong Kong-based contemporary potters resulted in a biased gender makeup within the potter community weighted towards women. Ms. B speculated that this could be due to the fact that men needed a more stable form of employment to support the family and pottery-making did not offer a lucrative full-time job option.

The sections above mostly discussed about aesthetics, design and ideas related to Ms. B's ideas on post-modernism. However, there is another side to her ideas about post-modernism as well – that of crafts-making, production-related issues and the technicalities of firing ceramics related to energy use and environmental awareness. Technology is a double-edged sword. On the one hand, as a symbol of modernity, it denotes progress, efficiency, advancement and the mechanization of means of production. On the other hand, technologies can be potentially disruptive to traditional lifestyles, the environment, production means and aesthetic outcomes. Here we talked about Ms. B's use of energy resources in making the products and the environmental awareness of that process. Like other Hong Kong-based potters, Ms. B used electricity for baking clay pieces in her studio kilns. The only chance when she had access to fossil fuel kilns was in Shigaraki, where she observed a pluralistic variety of production resources in the forms of electricity, gas, *raku*, charcoal, wood-fired kilns. The biggest difference in technique and methodology was that, in operating wood-fired kilns, the artist had to know when and how to close the air holes and insert the wood fuels which are procedures that are absent in electric kilns usage.

The skills for manual wood-firing require more than one-man-operations (OMOs) and need the cooperation of technicians, artisans and workers in the community. Ms. B describes Shigaraki as a potters' community with a production ecology serving the needs of the artists. For example, she mentioned that she was able to find a blacksmith in Shigaraki who could custom-make metallic bases and objects according to the specifications that she required. Ms. B noted that the same arrangements were not possible in Hong Kong due to lack of expertise and making the item in time for the show. Compared with Shigaraki, it was not as convenient to locate craftsmen who can do this kind of multimedia work to provide the required finishing component for her art pieces. In attempting traditional firing in the Shigaraki fossil fuel kilns, Ms. B pointed out that wood-firing kiln procedures required many days of preparations, shift duties, maintenance of kilns and hard labor to load it up with timber fuel. Ms. B suggested the reasons why Japan's pottery-making towns like Shigaraki remains operational are because of the scale of its industries, workshops and factories. The demand for ceramics products in Japan and overseas justified the presence of the division of labor in the potter communities. She opined that, due to the economy of scale, Japanese potters were able to survive economically despite keen competition from other pottery-making economies like China.

ENVIRONMENTAL AWARENESS

Neither Dirlik, Zhang and Wang featured the idea of environmental awareness dominantly in their narratives and ideas about post-modernism.[6] Given that a usual critique of modernization theories had been their excessive focus on quantitative output and standardization of mass produced items, I asked Ms. B about the importance of energy use and environmental concerns in her ideas about post-modernism and production, given that modern ceramics industry are producing large quantities of porcelain wares. In terms of environmental awareness in Hong Kong, Ms. B, like her peers, are aware of the regulations and impracticalities of wood-fired kiln work in Hong Kong. Therefore, Ms. B like many other potters in Hong Kong relies on electric kilns, probably the cleanest economical option available for the contemporary potter anywhere in the world. Other examples of environmental awareness cited by Ms. B encountered in her work include the avoidance of barium, a toxic substance found in some ingredients used for pottery-making. She would also only carry out salt-firing outside Hong Kong as it produces toxic chlorine gas. In Hong Kong, Ms. B hand-mixes the glaze for her works to keep track of the ingredients that go into making the glaze. Ms. B's ideas about protecting the environment in Hong Kong also have an element of nostalgia. Her ideas are influenced by distant memories of Hong Kong in the past that is less industrialized. She recalls earlier times when the city was not too urbanized or affected by pollution and one was able to view the mountainous landscapes in the city. It helped, therefore, that she now lives in the countryside with a modest garden plot where she can be in touch with nature and feel peaceful and calm to concentrate on her artistic works.

It is Ms. B's ideas about environmentalism in China that are most illuminating. Here, she distinguishes between the highly industrialized Jingdezhen and Yixing which she describes as a location that is environmentally-pristine and where artisans continue to pay careful attention to individual craft-making. It was in Yixing that Ms. B spotted folk pieces being made. Yixing has maintained its status as a representative ceramic art-form in China, specializing in purple clay teapots that distinguish its products from those of Jingdezhen. Jingdezhen as a pottery-making center in the world has attained a reputation for mass manufacturing fine porcelains for the global marketplace. In contrast to Yixing's clean environment, the image of a child who ate a piece of bread that dropped onto the ceramics dust found on the floor of a pottery-making factory in Jingdezhen made a lasting impression on Ms. B's environmental awareness. For Ms. B, it highlighted the dangers of excessive focus on profit-making goals to the exclusion of ethical production. Another example that Ms. B cited was the dim lights found in another Jingdezhen factory manufacturing realistic life-sized honeybee models with lifelike tiny legs. In Ms. B's view, poor lighting was endangering the eye sight of young workers in the factory. She was also conscious of the toxic gold and silver lusters (applied as coloring onto the bee-shaped products) found in these facilities.

In contrast, her experience in Yixing tells a different story. From her tour of the facilities, the workshop spaces in Yixing have neat and spacious tabletops and good lighting facilities. Yixing and Jingdezhen have different strategies when it comes to making profits and generating revenues. The higher-priced Yixing teapots that are individually crafted enjoy expert evaluations of higher artistic quality by their collectors. This is different from the mass-produced Jingdezhen porcelain cups which needed the economy of scale to make the same profit margins. By comparison, Yixing teapot-making requires the artisans to beat the clay slowly, then carefully carve out the teapot design and form. The manpower hours and individual efforts that go into making authentic Yixing teapots cannot be replicated by mass production.

ENERGY USE

Ms. B felt that energy use has a direct correlation with aesthetics. While electric kilns are convenient to use, she pointed out the danger that potters may have overreliance on technology and neglect to learn the skills needed for traditional firing. It also takes away the pleasure of accidental glazes by becoming too predictable in firing results. Due to the accurate controls of an electric kiln, Ms. B felt that she is not maximizing her cognitive problem-solving skills or making any physical exertion to meet any challenges that emerge in the firing process. Ms. B in fact had to resort to gas-kiln firing in Shigaraki to discover the artistic joys of manual reduction firing that result in unique pieces that cannot be reproduced in a repeated firing. In her decision-making process during a firing session, Ms. B does not place the factors of the convenience and affordability offered by electric kiln firing over all other considerations. She stated that even though electric kiln offers the option of cost-saving by reducing firing time, she would not restrict the duration of firing just to save finances and would continue to fire ceramics according to the time needed for her to achieve the glaze based on her artistic needs.

CONCLUSION

My dialogues and interviews with the Hong Kong contemporary potters revealed that they were reluctant to identify themselves as post-modernists. Some of them revealed that their artistic creations were not based on ideologies or any -isms. My contact with them revealed a community that was connected informally through personal networking and friendships. They did not exhibit any ideological ideas about capitalism, consumption or materialism but rather preferred to update each other on arts-related events, exchange personal articles like furniture pieces or souvenirs from travels overseas and share gossip. Their critique of capitalism is restricted to casual conversations and critiques of street talk on the cost of living, rentals and making ends meet. In other words, they were neither detached from practical economic concerns nor had strong ideological conceptions of capitalism. These observations appear to coincide with Wang's analysis that, while it may be possible while Hong Kong may not have a coherent social movement promoting folk art, there are pockets of communities driven and interested in producing and preserving traditional art forms, not necessary ceramics-making but a wide variety of folk arts and traditional desserts. These communities can be found in the districts of Sai Kung, Lamma Island, Stanley market and various other outlying and/or rustic villages and areas found off-the-beaten track in Hong Kong. In the case of the potter communities, they are found in bohemian communities ironically in industrial estates (symbols of industrial production) partly converted into artistic enclaves like Chai Wan and Fo Tan.

But the question is whether these districts are considered as holdouts from modernity or postmodern communities? They are likely to be an ambiguous mixture of both societies with residents keen to preserve traditional culture as a counter-reaction against mainstream mass consumption activities in Hong

Kong's cities and towns while enjoying the fact that infrastructural development had not fully reached some of these peripheral outlying districts. There are also artists who are keen to create their works in the company of fellow artists and in low-rent locations. This is yet another situation that pragmatically negotiates between costs concerns and aesthetic production. In other words, the artists as a group vacillate between conscious rejection of consumption culture and enjoying modern amenities on the other hand. They make conscious decisions in leaving the central business district area, situating themselves in less glamorous industrial estates, and enjoying a rural lifestyle in rustic villages.

Given the pragmatic impulses of the informants, it is difficult to determine if Ms. B is a non-ideological voluntary follower of Hong Kong's rural and traditional lifestyles, a post-modern rejecter of mass consumption and excessive profit-making, a bohemian artist with desires for self-actualization beyond what urban environments can offer or all of the above. The decisions she makes may rest on pragmatic lifestyle choices rather than ideological consciousness of post-modernist values. The non-ideological nature of the contemporary Hong Kong's potters' responses to the author correlates with Wang's observations as well. They are not an organized movement to reject conspicuous consumption in the city but are merely interested to lead lives according to their own pace and interests while consciously staying away from the rat race to pursue their artistic interests. To me, it is a form of pragmatic individualism, neither uncompromisingly dogmatic in beliefs nor unconditionally participating in capitalism. In her chapter, Wang appears to hint at the non-ideological nature of Hong Kong ideas about postmodernism and noted the highly pragmatic ways of how Hong Kongers conceptualize culture and traditions.

> There is, in Hong Kong, a clear distinction between the public and the private in terms of how people deal with their cultural differences. People are left to pursue their own way of life according to own cultural and ethnic tradition, as long as this does not stand in the way of the activities of the market.[7]

The informal nature of networking and establishment of trust based on friendship rather than institutions also finds correlations in Wang's work. In terms of organization of small businesses and cottage industries, Wang argues that, while the small businesses showed a tendency to depend on family members and relatives to operate the businesses, it was not based on kinship ties and values but a utilitarian arrangement to employ trustable employees.[8] Most Hong Kong contemporary potters that the author came into contact with were micro-entrepreneurs. They work on commissioned pieces for commercial outlets like hotels, ran classes for pottery-making students and ran retail outlets. Most of the partners and associates they worked with came from this informal network of friends, relatives and family members. They are merely contented to make their own crafts with minimal interference from profit-making concerns.

NOTES

1. This chapter is funded by the South China Programme Research Grant.
2. Xiaoying Wang, "Hong Kong, China, and the Question of Postcoloniality," in *Postmodernism and China*, ed. Arif Dirlik and Xudong Zhang (Durham: Duke University Press, 2000), 94.
3. Yee Fun Mak, "About Me," accessed October 28, 2012, http://makyeefun.com/about-me.html.
4. Ibid.
5. For more on this, the author suggests reading: Edmund de Wall, "Lucie Rie: Modernist Potter by Emmanuel Cooper," *Guardian*, May 25, 2012, accessed October 20, 2012, http://www.guardian.co.uk/books/2012/may/25/lucie-rie-modernist-potter-emmanuel-cooper.
6. Wang, "Hong Kong, China, and the Question of Postcoloniality."
7. Ibid., 101.
8. Ibid., 99.

Chapter Seven

Negotiating the Narrative of the "Last Frontier"

A Case Study of a Fourth-Generation Hong Kong Potter and His Art as Well as Environmental and Ethical Consciousness through Non-Governmental Initiatives in Myanmar (Burma)[1]

The past two chapters examined the reasons behind the choice of kiln technology in the case study of Hong Kong potters coping with ethical concerns about the environment, energy use, post-modern trends in consumption, constraints of urban spaces and the economics of firing ceramics. Following from these points, this chapter examines three main issues. First, I was keen to look at a case study of a Hong Kong professional artist who successfully integrated environmental and ethical concerns into his works. The term "environment" is used broadly in this chapter to refer to a combination of preservation of local cultures and the natural settings in which these cultures are located.

Second, I was interested in examining the sources of inspirations for Hong Kong artists coming from Myanmar (Burma)[2] and vice versa, resulting in the Myanmar-inspired series of works covered in this chapter. Amongst the Hong Kong potters involved in this series, fourth-generation ceramicist Mr. X's encounters with Japanese pottery traditions during his stay in Shigaraki became an inspiration for him to conserve such traditions in his works. His Japanese experience came from joining an artist-in-residence program in Shigaraki, Japan, where he realized the importance of pottery-making traditions. Several items in Shigaraki left deep impressions on him, including the conservation of wood-fired kilns for manufacturing ceramics using traditional techniques, despite the availability of newer electric kiln technologies. According to Mr. X, he could not find an equivalent infrastructure in Hong Kong with the same heritage objectives. Mr. X's own studio facility is a thoroughly contemporary studio equipped with electric kiln cramped into a small but cozy artistic space located in a building that sits in the concrete jungle surrounded by skyscrapers and other urban infrastructure in the central business district. For Mr. X to seek inspirations from natural environment for his works, other than the trekking the mountainous uninhabited terrains on the fringes of Hong Kong and Kowloon, Mr. X has to travel outside of Hong Kong to learn about other rural cultures and lifestyles. In this case, he went to Myanmar for his latest show at the point of this writing. In contrast to the urban jungle of Hong Kong's environment, Myanmar is described as the "last frontier," an oasis untouched by the fast economic growth that characterizes the East Asian "miracle" mentioned by the World Bank in the mid-1990s.

Third, in this chapter, I was interested to find out how artists in Hong Kong participate in non-state initiatives to look at ethical issues examined through artistic activities such as pottery-making. In this project, I am looking at the above research questions from a particularistic viewpoint—how observation studies of a single artistic venture (by a prominent fourth-generation Hong Kong ceramicist, Mr. X) can reveal some relevant perspectives in response to the issues raised above. I also conducted observation studies of Mr. X's inaugural show inspired by Myanmar titled "Burmese Days—A Timeline," held on September 13, 2012. This show brought together art enthusiasts, ceramicists based in global locations as far away as Yunnan, curators, potters, collectors, and other interested parties. Its audience outreach represents a slice of cosmopolitan Hong Kong society and its art scene—both the avant-garde art lovers and members of the public. The chapter does not pretend to be a comprehensive analysis of ideas and imaginations of "Myanmar" in the artistic and creative minds of Hong Kong's contemporary artists.

Together with his associates, Mr. X made a trip to Myanmar's Bagan (formerly known as Pagan) region. The mission statement and objective of the trip was clearly spelled out in the pamphlets distributed freely at the show: "We were in awe of the traditional crafts and arts. The constant threat to their existence with the recent opening up of the country makes the efforts to preserve handicrafts most meaningful."[3] The narrative in the handouts and brochures distributed by SOIL and Mr. X, was also straightforward and simple: "Spring 2012. Mr. X and two of his associates joined the SOIL team in an expedition to Bagan, the ancient capital of Burma, which has been closed from the outside world for almost half a century."[4] While dramatic in tone, it was indicative of expectations (from the adventurers' perspective) of exoticism, discovery, and ideas about untouched natural environments. SOIL is a non-government organization (NGO) based in Hong Kong which brings individuals together to promote the understanding of the history, culture, individuals and humanity that goes into making a piece of art and craft.[5] The stated objective of the SOIL group is that it "offers a platform for people to come together and learn about the history, culture, and humanity behind a piece of handicraft."[6] This statement provided hints of ethical production, typically associated with issues such as avoidance of destruction of traditional ways of life, advocacy of environmentally friendly production, etc. In an entry on its Facebook page, SOIL uploaded a message that characterized Burma as "a place to find your soul – pagodas connect between heaven and earth... So many things to learn in life, between lives, and in the process we experience, gain, and grow."[7] This Facebook entry suggests the existence of traditional notions of spirituality behind crafts-making in Myanmar. Mr. X himself did not elaborate much on this aspect of his works.

Combining the above narratives, three points appear to stand out. First, the Bagan trip represented a voyage of discovery. The artists involved had never been to Myanmar and they did not know what to expect in terms of inspirations and ideas offered by the traditional Burmese art forms. They did not know what to expect in the cultural scene of a society that was formerly isolated from the rest of the world, from modernity, and from the information and digital age. Therefore, they started off with an unknown in terms of what to expect. Second, there is a clear intention behind the social mobilization of like-minded artists and individuals to come together for this venture. The purpose was to learn about the history, culture, and traditions of another civilization or society in order to locate their own identities. By visiting what they considered as the marginalized fringes of the international society, the adventurers were able to locate their own unique identities and define the Hong Kong artistic style. Third, there was a certain degree of individual soul-searching and spirituality involved in this venture. This is the spiritual component of the trip where artists rediscover themselves by confronting their life priorities and challenges. Overall, there is a combination of cognitive, emotive, and spiritual elements involved in this Myanmar venture.

The SOIL group was started by potters Susanna Pang Bik-tsui and Venus Lee Lan-kiu, who sought to gain more recognition for their crafts, feature traditional handicraft works drawn from international sources, and organize fellow ceramicists with the same aspirations into a formal grouping.[8] SOIL eventually found a treasure trove of disappearing traditional crafts like lacquer ware in Bagan, Myanmar, and set up a store in Wan Chai to sell and distribute such traditional crafts.[9] This function appears to be a typical feature of non-governmental and non-profit groups trying to generate revenues to keep their activities financially self-sustaining. The group represents an eclectic set of ideas about non-state initiatives and loosely organizes their activities around ideas of traditional handicraft conservation and preserving traditional ways of life. The purpose is not to aggressively push in-your-face political or social advocacy messages, but rather to promote a softer approach of moral suasion to convince its audience of the need to recognize the importance of preserving the heritage of traditional crafts all over the world (especially in indigenous rural societies) and the protection of indigenous environments and habitats while promoting sustainable economic development for the national economy. SOIL and their messages do not appear to be ideological, but based on care and concern for fellow artists around the world. The following tract is an example of SOIL's referral to other works, featured on SOIL's official Facebook page:

> Burma is a country blessed with rich natural resources, traditional crafts, and cultural diversity, but it is cursed with many problems...Shelby Tucker, the author of this book, traces the impact of colonization, as the country was eventually torn apart by the seismic events of the Japanese invasion and British re-conquest. Following the killing of Aung San and the British departure, this book also analyzes the downward spiral of Burmese misrule, its mistreatment of minorities, and cooperation with international traffic in drugs ... This is a must-read for anyone wishing to catch up on Burma's past and to understand Burma.[10]

The online passage recommends readers to acquire books with contents that focus on more detailed Burmese studies, indicating that their main interest is not really in public ideological and intellectual debates, but rather a public appeal to interested parties to learn more about the history of Myanmar. SOIL's show on Myanmar that I attended and its website are not textually prolific, instead both platforms rely on visual presentation to stimulate individual interpretations of the subject matter. The ambiguity of art forms is sometimes more effective than literal words in putting across subtle political messages and slogans. The softer approach may also be more persuasive for some audiences who are less accustomed to more upfront forms of activism. I based my observation studies on a SOIL event held on September 13, 2012 where I interacted with Mr. X and other event organizers. An online search reveals that SOIL had previously organized another related event centered on Myanmar titled "Hong Kong Creative Art Explorers Series: Understanding Burmese Lacquerware," which featured talks on Baganese lacquer ware, complete with a multimedia visual presentation accompanied by wine and cheese for audience members.[11] The workshop's website featured the following quotation on website publicity: "Let handmade crafts take you into a lesser known culture to meet with the people behind the creations, to understand the process, and to appreciate the skills."[12]

This website blurb was indicative of two points. First, it emphasizes the idea of traditional crafts made through manual labor and associates it with minority indigenous cultures. There are a number of slogans that emphasize manual efforts in making handicrafts on SOIL's official Facebook page: "Soil offers a personal and unique perspective on what our hands can create."[13] The organization associates its art preservation efforts with an emphasis on creating art works using bare hands to better understand the traditional process of art creation as well as its challenges. SOIL's collaborative partner, YY9 Gallery, is a cozy bohemian gallery in the upper-income district of Happy Valley. The gallery space consists of two sections, one designed in the form of a tasteful, artistically-conceptualized narrow corridor space and another space for displaying works for sale. The gallery owner revealed that he was acquiring additional properties in Chai Wan (a nearby industrial estate with a thriving community of artists working on a variety of art forms).

In addition to working with galleries, SOIL's collaborative spirit with other stakeholders in Hong Kong's art scene can also be detected through the publicity created for their Wan Chai pop-up store featured in *Timeout* lifestyle/entertainment magazine and the Hong Kong Tourism Board website's section entitled *Discover Hong Kong*. The shop features East-West fusion art works by Western artists (like British-Ukrainian Veronica Gritsenko) who experimentally hybridized traditional Myanmar art with contemporary Western forms.[14] Conversations with Mr. X revealed that East-West fusion and hybridized art forms have mixed reviews in Myanmar, where they are received with an ambiguous combination of resistance and acceptance. In this sense, SOIL is promoting cutting-edge art forms that are not yet universally accepted, and this comes with a certain degree of risk as responses (and acquisitions) from Hong Kong and international buyers are unknown. In this way, SOIL showcases innovative artworks to the public at large even before their complete acceptance by the artistic community.

Mr. X's works represent this cutting-edge innovative spirit. He is not advocating the preservation of traditional handicraft artwork by simply acquiring existing pieces. He reinvents those works by reinterpreting traditional styles and reconstructing traditions in modern art forms representing rather unique local Hong Kong adaptations of traditional art works found in Myanmar. The purpose is to stimulate awareness amongst art lovers of the significance of preservation of traditions and culture amongst art patrons and individuals who have come to appreciate these works. The sense of loss of traditional art and a desire to collect them motivated Mr. X to collect novelty pottery mugs made in wood-fired kilns in Japan that cannot be reproduced by electric kilns in urban settings. My informants tell me that existing rules and regulations forbid local potters from setting up wood-fired kilns in Hong Kong. Mr. X also collects Qing kitchen pieces, unwanted artifacts from old shops with artistic value and ceramics with glazes made by other Hong Kong potters that he personally cannot reproduce. In this role, he is a preserver and curator of traditional arts.

Similarly, the same motivation drove Mr. X and his associates to display lacquer wares they collected during the Bagan trip at their inaugural show on Myanmar, in addition to their own works made in the same location. In this sense, the artist in this case study is not only collecting traditional art from villages, but also remaking them in his own interpretative styles to stimulate audiences to ponder over both artistic and developmental issues in Myanmar. Part of this chapter is also based on psychoanalyzing Mr. X as an artist to understand his role in grappling with ethical concerns related to his immediate artistic environment. Aside from his Myanmar-associated works, Mr. X has been concerned about the destruction of the environment by toxic glazes, poisonous fumes from burning of used wooden furniture pieces, and recycling wastes from unused clay. Perspectives from the artist himself are representative of the worldview held by the contemporary fourth-generation group of Hong Kong artists who are just reaching the peak of their artistic careers. Though the case study in this chapter may not be representative of all artists, it provides a useful snapshot of some common issues faced by contemporary artists in Hong Kong.

"THE LAST FRONTIER"

In constructing the literature review for this chapter, I was interested in a Mekong-centered perspective of disappearing art forms in that region. Modernization and urbanization tend to be cited as the main causes of disappearance of traditional ways of life. In Myanmar's case, there is an additional angle to this phenomenon. At the time of this writing, Myanmar was touted as the "last frontier" of development in East Asia. Conversely, this narrative of Myanmar also means that the imagery of Myanmar as a state and economy represents the last bastion of traditionalism found in the East Asian landscape that is relatively untouched by foreign direct investments (FDIs), economic development, and commercialization. Mr. X mentioned that the Myanmar districts they visited were only just welcoming some of the global fast food restaurant chains into the country. Media headlines in publications such as *Forbes* carried headlines describing Myanmar as the "last frontier" of development in East Asia. *Forbes* was probably one of the first global media outlets to dub Myanmar as the "last frontier"[15] of Southeast Asia. Myanmar's democratization and economic opening-up and reforms are likely to be highly positive for its economic development after years of junta-ruled political systems. In fact, there is a sufficiently long history of multinational companies operating in Myanmar, including the likes of France's Total, China's China National Offshore Oil Corporation (CNOOC) listed in Hong Kong,

South Korea's POSCO (Daewoo International unit), and Thailand's PTT.[16] In this list, Hong Kong's listed company is visible as one of the earliest multinationals investing in Myanmar.

Therefore, Mr. X and his associates are entering Myanmar at an interesting juncture as the economy of that country transitions from a closed military junta-managed economy to a young market economy that is opening up. Born and raised in Hong Kong, Mr. X fully understands how economic development can challenge traditional lifestyles. After all, Hong Kong was one of the four tiger or dragon economies that developed rapidly after Japan in the postwar years in the so-called East Asian economic "miracle" period. Rapid economic development made Hong Kong the international financial center that it is today and also created a metropolis dotted with skyscrapers and attained one of the world's highest GDP per capita. The economic benefits of prosperity are clear. But such prosperity also led to the alienation of traditional arts and crafts. For example, on Peng Chau Island, the Lams are the only remaining potter family left on the island still making local pottery after the golden age of the local pottery-making industry in the 1960s when a number of families were engaged in the trade. The rest were outcompeted by cheaper imported wares or lost their competitiveness due to the increasing wages in Hong Kong that came along with economic prosperity.

Another change influenced by economic development is the relocation of artists' studios and residences to the peripheries of Hong Kong. Many contemporary artists made conscious decisions to relocate away from the prosperous urban centers. According to my informants, some artists like third-generation Hong Kong potter Mr. A moved to the outskirts of the city (Sai Kung) to pursue their artistic ventures. The exact reason for relocating to the margins of Hong Kong is not precisely understood. With Hong Kong's disappearing pottery-making scene due to economic progress, Mr. X is also cognizant of the fact that Myanmar folk artisans may no longer be interested in pursuing traditional craft work due to at least three similar factors. First, the increasing mechanization of industries in Myanmar means that making traditional crafts will give way to efficient and standardized mass-manufactured products made possible with the importation of new machinery from foreign sources. Second, lucrative jobs and salaries might increasingly attract younger artisans, who then give up family traditions or learning traditional skills altogether. Third, economic development might also change the natural landscapes of rural areas and large tracts of lands where natural raw materials and ingredients are found, leading to reliance on chemical compounds.

Mr. X revealed his inner feelings about the loss of lacquer ware traditions in Myanmar through his art works. Mr. X went through a rough period in his artistic career just before he became an artist-in-residence in Wisconsin and he expressed his angst and frustrations through installation pieces created at that point of time, featuring broken porcelain limbs, pottery wrenches, and ceramic guns. Since then, he has come to peace within himself and the angst that he had is minimized in the Myanmar-inspired pieces and replaced by a sense of sympathy for the loss of traditional art forms and incorporated his ideas about development and environmental preservation. Mr. X's associates his life journey with pottery-making activities: "Time is a river. It will not wait and everything depends how you live with it. Through clay, I find my way [Mr. X]."[17] This quotation appears to suggest that Mr. X intends to continue with his life journey of discovery in pottery-making, regardless of obstacles or challenges ahead. He plans to live life to his fullest. Having confronted his inner feelings about personal life matters, Mr. X seems to have found a new goal in artistic ventures – to interpret traditional arts and emphasize its contemporary importance while making efforts to gather and preserve old traditions through collecting artifacts, observing traditions, and innovating contemporary pieces.

Therefore, Mr. X's motivation for artwork creation is based on a cognitive awareness of what development may bring to Myanmar (Burma). Mr. X's awareness is not grounded in scholarly discourses or economic theories, but in personal observations. His artistic quest is the rawest expression of this awareness and understanding of the changes facing Myanmar, uncomplicated by elaborate theoretical ideas. By comparison, the SOIL group is relatively more proactive in outreach by distributing pictures of natural sceneries found in Myanmar (Burma) such as the Irrawaddy River of Bagan Myanmar (Burma), color-coded maps of Myanmar (Burma), scenes of traditional Myanmar (Burma), photos of religious sites, pictures of traditional handicrafts, and paper boats with traditional Burmese motifs.

The current ongoing economic liberalization of Myanmar has not yet been vigorously discussed in academic writings because of its relatively recent occurrence. However, the same issue was vigorously debated in international media commentaries[18] due to the somewhat unexpectedly rapid pace of political reforms in that country after decades of military rule. Foremost on the list of attractions for investors are Myanmar's abundant natural resources and its strategic position as a pivot between the two large emerging markets of India and China. Plentiful natural resources, abundant affordable labor, a relatively young workforce, and untapped consumer markets are all attractive points for international investors.

GLOCALIZATION

The process of "glocalization" appears to be present in Mr. X's work. His work represents a local Hong Kong interpretation of Myanmar art forms. Glocalization represents a hybrid model of globalization combining advancements in transportation, communications, and digitization of information with local interpretations of global affairs, trends, and events. In underlining the disappearing traditional arts of Myanmar, Mr. X places greater emphasis on the element of local traditions rather than the global context. For example, in the untitled piece below, Mr. X revealed that the gold color of his work (middle piece) was inspired by the gold leaves on Burmese pagodas and the red clay color (right piece) reflects the color of the lacquer wares of Bagan.[19] Mr. X admitted that the piece also reflects his concerns about the disappearing traditional arts of Myanmar due to economic development that came along with Myanmar's opening-up.

The focus on Bagan handicrafts is an important one. Bagan is a major lacquerware-producing center in Myanmar. I surveyed academic literature on Bagan lacquer ware and discovered a rich collection of writings on this subject matter. The ingredient that makes Bagan lacquerware unique and visually stunning is the varnish that is applied onto the handicraft products. George Watt wrote a seminal article on this Burmese varnish (*Melanorrhoea usitata* or Thitsi). According to Watt's article, a special characteristic of this paint is that the varnish becomes dry and sets most effectively in a humid environment, which is especially interesting since Bagan has the most arid weather among Myanmar's dry regions.[20]

In order to imitate this varnish texture and appearance, Mr. X admitted that he used toxic metallic oxide glazes to achieve the color. Therefore, Mr. X was careful to indicate to all admirers of the Burmese-inspired pieces that they were made exclusively for display only and not for use at the dinner table. The efforts that Mr. X puts into imitating Burmese-style lacquer ware colors underscore his determination to preserve the heritage of Myanmar's traditional handicraft sector accurately through local interpretations. In the same exhibition and show, Mr. X's associates made paper boats out of materials that resembled the luscious lacquer ware of Pagan.

The skills and effort needed in the application of the varnish paint onto Baganese wares is described succinctly by Watt:

> After being engraved the article is handed to another operator, who places it on the turning-lathe, and taking a small quantity of some dry metallic pigment in the hand, rubs it all over, and thus loads the engravings with the colour. The excess is rubbed off and the colour fixed by a coating of varnish…After being placed aside to dry, the article is handed back to the engraver, and other portions of the design are scratched or chiselled out and loaded with their special colours and thereafter varnished. Time after time this process is repeated until all the colours desired have been added and the design completed. The article is then given one or two final coatings of varnish.[21]

Mr. X's work is further complicated by the fact that he has to use his knowledge of chemistry in selecting the right glaze for his works in order to emulate the Baganese varnish color. It is a surgical attempt to use processed ingredients to create a color that comes inherently with a natural organic dye. Here, the achievements of pioneering generations of potters before Mr. X may be lauded. A second-generation potter, Mr. CC was prominent in the 1950s and 1960s for creating his own glazes and admired for his knowledge of chemistry. Mr. CC opened the way for other Hong Kong-based potters to develop their own glaze techniques, including his original "jade" glaze. Mr. CC discovered jadeite stones in the rural district of Sai Kung and crushed them into powder form before integrating them into the ingredients to create a glaze for application on his pottery works. Mr. CC also inspired later generations of contemporary Hong Kong potters to acquire knowledge of glaze chemistry and firing techniques.

AESTHETIC SPIRITUALITY

Aesthetically, there is an element of spirituality in Mr. X's pieces, particularly the centerpiece of the inaugural show on Myanmar which is a pagoda-inspired installation work touching the ceiling of the gallery. It symbolizes a ceramic structure extension into the skies, a connecting bridge between heaven and earth. The work reflects the traditional beliefs of Myanmarese villagers in linking up the two realms of existence. The idea of spirituality is also written in the show's pamphlets: "Buddhism has deeply permeated Burmese life, manners, and aspirations, despite the continued territorial disputes and ethnic conflicts over the years."[22]

CONCLUDING SECTION

From interpreting his works, I get the sense that economic development is just as important as ideas about cultural and environmental preservation, and probably for the people of Myanmar as well. The central message that Mr. X was conveying to the public through his exhibition "Burmese Days – A Timeline" is the need for balance – reaching an equilibrium between developmental priorities and conservation of the environment as well as indigenous lifestyles. With the benefit of hindsight after rapid economic development in Hong Kong, Mr. X projects his ideas about the excesses of urbanization onto his art works, inspired by the traditional lifestyles of a society on the verge of market-driven economic development. Mr. X also practices what he preaches. He recycles materials in his workshop, has a fascination with green plants, and constantly reminds his students about minimizing wastes and avoiding toxic glazes when making table wares. The combination of tradition and modernity is symbolized in Mr. X's works on at least three levels: modern interpretation of a traditional art form, firing of Baganese-inspired pieces using a combination of manual labor and modern electric kilns, and emulation of traditional colors by using artificial chemical compounds.

I presented the contents of this chapter at an international conference on Myanmar and non-governmental organizations (NGOs) organized by The Chinese University of Hong Kong in October 2012. I was gratified to receive positive comments about the content. The discussions contributed to a gap in the conference literature, namely, the presence of fine arts as a platform for NGO networking. Given the attention paid to NGOs by the experts at that conference as well as the popular media, the topical matter of the fine arts scene changing in Myanmar, which is not related to the issues of political reforms, governance, economic opening and integration of opposition politicians, was refreshing for the participants. While it is significant that political communities are being formed through the dynamics of the democratic system, communities of artists are also coming together.

These communities can be in the form of physical networks that connect traditional crafts with fusion art works that combine modernity with traditions or a result of East-West hybridization. Communities can also be imagined, binding together artists and craftsmen with eclectic ideas of heritage preservation and conservation of traditional arts and culture. This collection of vague and ambiguous ideas about heritage and traditional arts may not exactly coincide and may even be dichotomous, but they serve as an overarching framework for discussions among interested stakeholders. In the constructivist sense, the network of ideas that bind these individuals together may have the potential of creating a common identity and value system around which future institutionalization may take root and grow. This also forms part of a democratic system characterized by pluralistic institutions in a free society. Political activism may bring about transformative social changes (the "hard" components of political reforms), but artistic expression can be a "soft" form of moral suasion in influencing social changes as well. Both components are needed for a truly democratic society to take root.

NOTES

1. The South China Programme Research Grant funded this chapter.

2. I follow the preference of the organizer of the conference in using "Myanmar" in this chapter.

3. Derived from an untitled handout at an untitled show in a gallery located in Happy Valley.

4. Ibid.

5. Please refer to the "About SOIL" section of SOIL's official website at: SOIL, "About SOIL," accessed August 2, 2013, http://craftbysoil.wordpress.com/about/.

6. Derived from an untitled handout at an untitled show in a gallery located in Happy Valley.

7. SOIL's Facebook page, September 16, 2012, accessed September 20, 2012, http://www.facebook.com/craftbysoil.

8. Vanessa Yung, "Myanmar Pops Up Here," *South China Morning Post* (SCMP), April 29, 2012, accessed September 20, 2012, http://www.scmp.com/article/999498/myanmar-pops-here.

9. Ibid.

10. SOIL, "Burma: The Curse of Independence by Shelby Tucker, 2001," Craft by Soil, September 9, 2012, accessed September 20, 2012, http://craftbysoil.wordpress.com/2012/09/09/burma-the-curse-of-independence-by-shelby-tucker-2001/.

11. Art Exploration Society, "Hong Kong Creative Art Explorers Series: Understanding Burmese Lacquerware," May 11, 2012, accessed September 20, 2012, http://www.meetup.com/Art-Exploration-Society-Hong-Kong/events/62875472/.

12. Ibid.

13. SOIL, "Burma."

14. Time Out Hong Kong, "Four Things We Love About...Sau Wa Fong," Discover Hong Kong, Hong Kong Tourism Board, May 23, 2012, accessed September 20, 2012, http://www.discoverhongkong.com/nz/shop/whats-hot/sau-wa-fong.jsp.

15. Simon Montlake, "Clinton Visit to Myanmar Stokes Interest in Asia's Last Frontier," *Forbes/Wall Street Journal*, November 30, 2011, accessed September 19, 2012, http://www.forbes.com/sites/simonmontlake/2011/11/30/clinton-visit-to-myanmar-stokes-interest-in-asias-last-frontier/.

16. Ibid.

17. Mr. X's new media quotation. Details removed here to retain anonymity.

18. For a typical representative example of an international media article highlighting optimistic narratives of Myanmar's economic development, please refer to *Boom Days in Burma* found in the bibliography of this chapter.

19. Readers can refer to this website for royalty-free photos depicting the pagodas of Bagan. Dreamstime, "Royalty Free Stock Photos: Ruin- Pagan, Burma (Myanmar)," accessed September 19, 2012, http://www.dreamstime.com/royalty-free-stock-photos-ruin-pagan-burma-(myanmar)-image9838658#.

20. George Watt, "Burmese Lacquerware and Burmese Varnish. (*Melanorrhoea usitata*, Wall)," *Bulletin of Miscellaneous Information (Royal Gardens, Kew)* 1906, no. 5 (1906): 142.

21. Ibid., 144.

22. Derived from an untitled handout at an untitled show in a gallery located in Happy Valley.

Chapter Eight

Shek Kip Mei's Artistic Green Village

The previous chapters discussed the different resources that are used for firing and how the choices of fuels have evolved according to the needs of the industry, environmental awareness and creative needs. I will now situate these ideas in the four comparative case studies of Arita, Jingdezhen, Yingge and Hong Kong's art clusters and examine their similarities and differences starting from this chapter[1] . This chapter focusing on Hong Kong's Jockey Club Creative Arts Center (JCCAC) is the first case study in the volume. Its unique multistory layout, urban setting, integration of green spaces into the design and the presence of various artists working in different media contrasts with more specialized pottery-making towns like Arita, Yingge and Jingdezhen that will be featured in the subsequent chapters. In December 2012, I was invited to shows that featured ceramic and multimedia works in the JCCAC where I carried out observation studies. The JCCAC is conveniently located in Kowloon, accessible by subway (nearest station Shek Kip Mei Exit C is a five minutes' walk) and is open daily for 12 hours (10 am to 10 pm). The institution is a fixture in Hong Kong's arts and crafts scene. The facility is a one-stop location with two culinary cafes, dance studio (known as the White Box DanceArt), workshops, schools that offer short-term art courses and information centers on university degree courses, etc. Artists specializing in different art-forms freely exchange ideas with each other, enhancing opportunities to create hybridized works that fuse East-West and tradition-modern influences in artworks. There is also a multidisciplinary artistic environment in the facility as it houses fine arts specialists, including visual artists, theatrical actors, performing artists and textile designers amongst others. The presence of an assortment of art forms encourages cross-collaborations across different art-forms. Ceramic art, for example, is combined with music as the Hong Kong Ocarina Cultural Association organizes ceramic flute music classes in the building. Here too, customers (gallery owners, collectors and members of the public) and producers (artists) can match their creative and acquisition needs with each other under one roof. Some gallery owners who are artists themselves let out spaces for other artists and friends to display their works. The atmosphere exudes community self-help. During my observation studies, I saw a gallery owner shifting furniture and personally cleaning out the litter from the launch party of the same event. The galleries are multifunctional and sometimes serve as workshop spaces where collaborative work is carried out. The artists, in return, get public space to display their works. The environment in JCCAC is a multitiered community space for multidisciplinary cultural interactions. This chapter is based on my observation studies of JCCAC (Hong Kong's first dedicated arts village). I also collected texts related to the arts scene in Hong Kong and JCCAC from public spaces like the airport, universities, malls and subway stations as well as JCCAC-related paraphernalia. I also conducted conversations and interviews with local artists.

BRIEF LITERATURE REVIEW: ART CLUSTERS, GENTRIFIED INDUSTRIAL DISTRICTS AND COMMUNITIES

Existing literatures about gentrification of former industrial districts transformed into vibrant art districts are included here. Many case studies of gentrified districts tend to be located in neighborhoods that fit the profile of a post-industrial economy in decline. In the case of JCCAC, Hong Kong, where it is located qualifies as a post-industrial service economy since many labor-intensive manufacturing industries shifted to the Pearl River Delta region, specializes in the global banking and finance industry. However, there are few areas in Hong Kong that is indefinitely and unequivocally in permanent decline because of the city-economy's shortage of space for a dense population. The demand for commercial space will persist as long as Hong Kong benefits from its hinterland economic connection with China, an effective economic integration into the global economy, continuance of its laissez faire capitalist system as well as maintaining its status as a commercial center for a dynamic Asia Pacific region. Spaces in Hong Kong, old and new, small or expansive, are constantly renewed, rebuilt, revitalized for various commercial and residential purposes. Due to the premium of space in Hong Kong, especially in areas like the bustling Kowloon Shek Kip Mei area where JCCAC is located, not many areas qualify as a permanently declining area by many measures. Instead, the description of "older district" with an aging community is probably more appropriate to describe Shek Kip Mei as an aging residential area in the territory. To understand Shek Kip Mei better, I turned to the housing authority's historical account of the area's development found in online resources. The construction of a housing estate in Shek Kip Mei was a consequence of systematic plans to build high-rise residential units after transitional housing (constructed in the aftermath of a 1953 great fire that destroyed squatter areas) proved inadequate.[2] In the 1990s, the authorities removed old resettlement units and redeveloped the area in four phases (6,600 flats for rent to accommodate 17,400 individuals) and completed the first phase in 2006 when two high-rise buildings were completed.[3] JCCAC's gentrification in the industrial sector complemented the overall renewal and revitalization program for Shek Kip Mei's residential neighborhoods although they may not necessarily belong to the same urban-planning master-plan. In the area of the environment, both residential and industrial unit gentrifications have visible green features (these are discussed later in the essay).

The process of gentrification is associated with idea of urban decay. In the case of JCCAC, the deteriorating condition of the former Shek Kip Mei industrial building symbolizes the visible and physical decay of the asset. John McCarthy distinguishes between two kinds of regeneration efforts of decaying areas. He quotes Evans in defining culture-led regeneration in which culturally-related events and activities are the main motivators of urban repair and regeneration.[4] He contrasts this with Evans' definition of cultural regeneration in which cultural activities are but a component of overall regeneration efforts.[5] In the case of JCCAC, the former definition (culture-led regeneration) is probably more applicable since it is a culture-led renovation of a former industrial estate. However, as an art community-led component that was coincidentally proximate to Shek Kip Mei's overall urban revitalization efforts, the latter definition becomes more applicable. JCCAC is a cultural component of the authorities' overall efforts to revitalize the district in general.

While residential units and public housing are relatively immune from economic downturns, industrial units in the area are subject to market-sensitive occupancies. During economic downturns, spare capacities in industrial parks in Hong Kong are utilized for artist tenants. According to Chiu (Director of General Administration HKBU and Member of the Governing Board of Hong Kong Creative Arts Centre Limited), bad external economic conditions prompted the idea of using spare capacities in industrial parks and districts for use as creative workshops.[6] Therefore, selected spare capacity in industrial estates sometimes act as buffer units for rent and sale at time when occupant tenancy is low. In this way, JCCAC and other industrial estates with artist clusters (e.g., Fo Tan and Chai Wan) qualify as gentrified districts. Rents in these gentrified sectors draw artists if the rates are affordable. For artists, the socioeconomic struggle to meet rental rates is persistent throughout their careers. Many artists need sustained periods free from financial worries and anxieties to concentrate their undivided artistic energies on creative and artistic works. Ley cites Bourdieu in defining artists as individuals with "high cultural capital and low economic capital."[7] In some ways, this resonated in the conversations with the artists whom I met at JCCAC on December 9, 2012 who mentioned that they appreciated the stable and slightly below market rental rates offered by the facility. With financial worries taken care of to a certain extent, they could then carry on with their creative production work. Gentrified sectors also offer splendid isolation in which artists are able to carry out their works in relatively quiet environments free of noise pollution, particularly in a densely-populated city with its high density of vehicles.

The JCCAC is not particularly quiet although it sits in a comparatively less boisterous area of Kowloon city. When I went to conduct observation fieldwork at this facility, the first cab driver that I flagged down claimed that he had no idea where JCCAC was located and requested that I dismount. The second taxi driver knew and ferried me to the location although he required some time to spot the actual building. On my second observation study session, I took a cab from Kowloon Tong station to JCCAC. This time, the taxi driver knew where Shek Kip Mei was but was rather unclear where Pak Tin Street was located. After reaching Shek Kip Mei district, through my directions, the driver eventually located JCCAC. In other words, not all cab drivers who ply the road on a daily basis knew where the location was. JCCAC does not fit the stereotypical images of Kowloon – which is typically imagined as a cluster of large malls, urban university settings, residential units and "old rich" high class residential area. Socioeconomically, the location of an art school in such settings does not fit un-problematically with the stereotypical view. By comparison, Fo Tan and Chai Wan artistic clusters are located conveniently near a subway station with mini buses plying the routes to reach those industrial estates. Ostentatious consumption associated with Kowloon's shopping areas is not obvious in the areas around JCCAC. Even within JCCAC, the artistic atmosphere imposes an unconscious check on over-commercialization at the entrance. The only units that possibly qualify as retail outlets are the row of shops on the lobby floor that sold handmade (not mass-produced) artistically-designed jewelry and crafts signifies creative production that de-emphasize mass production in favor of tasteful handmade products.

Walter Santagata associated the importance of arts and culture with economic benefits (creative production, intellectual concepts and specialized skills, visitors/tourists attraction); social legitimacy (intellectual property and contents, public recognition) and environmental value-added-ness (for both developed and emerging economies).[8] Are these features applicable to the JCCAC case study? Economically, JCCAC represents revitalization of decaying and decommissioned industrial space for use by productive artistic ventures. Its events attract visitors although I have not seen any materials used for tourist promotion (with the exception of publicity for the JCCAC festival in the pamphlet Artmap discussed below) that features JCCAC in my textual collection of arts-related material in Hong Kong. In terms of social legitimacy, JCCAC's association with Hong Kong Baptist University (HKBU), a public totalisator board and the fact that it was opened by the former chief executive indicated there are prominent stakeholders in its success. The vitality of the intellectual and creative assets in the building is clear from the events-packed schedule found in its brochures, pamphlets, social media publicity and other public dissemination of information about its activities. Besides richness in arts contents, the performances and activities are also environmentally-friendly, designed to generate minimal or to recycle wastes. I will discuss the environmental aspects (in terms of green features and respect for nature) in a later section below.

At the point of this writing, there is currently no existing literature written specifically on JCCAC that I was able to locate through the HK-based university libraries, word of mouth amongst the artist community and Google Scholar. This is despite the fact that JCCAC is Hong Kong's first dedicated artist village and a creation of large-scale funding from a public totalisator board. Nevertheless, there are some useful references written by stakeholders familiar with the JCCAC institution. For example, Chiu's power-point presentation at the *Half Day Symposium on Revitalisation of Factory and Old Buildings* on January 22, 2010 is a valuable reference source on the making of JCCAC.[9] The significance of the Hong Kong case study lies in the fact that it is part of the creative industries push (intentionally or otherwise) from the time the initiative appeared in the Chief Executive's Policy Address in 2003 (in comparison, Desmond Hui points out creative industrial projects only made debut in mainland China at the start of 2005).[10] In other words, Hong Kong is at the forefront of the creative industrial economy within China. Hong Kong's progress may therefore serve as a sort of precedent for the rest of China. And JCCAC as an artistic creative production center is a component of Hong Kong's creative industry push, it is also a pioneer in this area as Hong Kong's first multi-tiered dedicated arts village. The areas of the creative industries that the Hong Kong authority defined as priority areas are "film, television, music, design, architecture, comics and animation, games and digital entertainment, etc."[11] Publicity for these industries can be seen in the public spaces. The activities of JCCAC falls into the areas of film, television, music, design, comics and animation, in other words, most of the areas that the government had identified.

As a case study, JCCAC is not overtly associated with any deep socioeconomic problems and challenges. Other than the unsatisfactory conditions of makeshift public housing built after the great fire in the 1950s in Shek Kip Mei, the 1990s attempt to refurbish residential units in the area as well as the early 21st century decommissioning of the Shek Kip Mei industrial estate

(JCCAC's predecessor), I was unable to find any materials on socioeconomic tensions, organized popular resistance (other than local residents' complaints of excessively bright lighting in JCCAC mentioned later in the essay), state-level-local governmental political dynamics or crime infestation that plagued many other decaying or old neighborhoods in prominent case studies on gentrified areas found in existing literature. It does not have the problems and challenges found in the works listed in the bibliography such as: tensions over control of neighborhood gentrifications between the private sector and the state bureaucracy depicted by Jason Hackworth and Neil Smith's study of districts New York city; the complexities of central planning found in Desmond Hui's work; the re-positioning of socialist images in Jennifer Currier's work, the anxieties of urgent employment creation found in Walter Santagata's writing, or issues of racism in Valerie Kinloch's study of Harlem art community.[12]

In the case of JCCAC, a totalisator board works closely with the government and a university along with other consultative groups and stakeholders for the initiative. There are no obvious fissures between state and local capitalists in the venture. Because the Special Administrative Region (SAR) government works with a tertiary institution and other community bodies, there are no strong central-planning currents found in Hui's writing. Given the SAR's market-oriented economic development, it does not have any socialist legacies to contend with. The rate of unemployment in the territory currently does not qualify as a symptom of a major economic decline. Nor has Hong Kong experience tremendous challenges like that of the severe acute respiratory syndrome (SARS) infection in 2002 or the economic crisis in 1997. No data is available to see if JCCAC's origins or operational management were related to these events.

Much of the textual information that I received focused on physical infrastructure renewal (water leakage and concrete deterioration)[13] as the major reason for gentrification rather than unemployment arising from the closing down of Shek Kip Mei industrial building. I did receive feedback from an artist that, in the beginning phases of JCCAC's foundation, there were disagreements within the arts community about the sort of artistic activities that should flourish in the building as well as its physical layout (for example, some preferred open spaces rather than individually enclosed units. These disagreements do not represent any deep structural socioeconomic schisms or divisions in society and are common expressions of personal opinions about art village and creative production facilities in any setting.

In discussing the idea and concept of cultural heritage, Luciana Lazzaretti classified major scholarly works in this area into three groups:[14] (1) those related to the contribution of intellectual and cognitive elements in an information society with highly-developed communication technologies crucial for a city's development; (2) studies evaluating the ability of local districts and enterprises in producing globally-competitive products for art consumption; and (3) research work concerning the ability of an economy to produce art works with unique identities based on locally available resources. Hong Kong as a creative city probably meets all three criteria, but what about the case study relevant to this paper, how does JCCAC's role fit into the cultural heritage concept? In the case of JCCAC, the intellectual resources that the building possesses are enhanced by the cosmopolitan resources available to Hong Kong as an international business city. Artists and designers from different parts of the world have the opportunity to collaborate with local artists who are themselves exposed to global art influences in creating unique fusion and hybridized works through innovation and re-interpretation of traditional arts (For example, both Ms. A and Ms. B, informants for this study, were artists-in-residence in Shigaraki, producing unique works of art that fuse traditionalism with innovative ideas). The JCCAC also hosts art events whereby artists can participate in bazaars to display their works to members of the public. The fundamental idea behind JCCAC's stable rental rates as an incentive for hosting art events is to incubate new artistic talents and encourage the development of the fine arts industry. The use of recycled materials made from products left over from the building's industrial production in the past to create art works is a unique way of preserving heritage, combining Hong Kong's economic heritage with arts consumption. The fusion of industrial history, fine arts creations and cultural consumption encourages fine arts production rooted in local identities (artistic renditions of products that the Shek Kip Mei industrial building used to manufacture in the past). In the same way, the display of indigenously-made machine tools as fixtures in the building preserves industrial artifacts as cultural heritage. Consequently, following Evans' writing, JCCAC itself is a culture-led regeneration and not cultural regeneration.[15]

HISTORY OF JCCAC

In his power-point presentation delivered on January 22, 2010, Chiu described JCCAC as Hong Kong's first artist village and profiled it statistically: the facility contained 124 artistic and creative units covering a total area of 110,000 square feet.[16] According to a local heritage group Hulu Culture, the JCCAC building was finished in October 1977 at a time when industrial areas were segregated from residential units in the Shek Kip Mei district but became disused in 2000.[17] But according to the pamphlet *Heritage Fiesta 2012* published by organizer Commissioner for Heritage's Office Development Bureau, the Shek Kip Mei Industrial Estate was a family's private property built in the 1960s that was later opened to the public in 2008 as an arts village.[18] The involvement of the Heritage Office indicated the significance of the asset as a symbol of increasing public awareness of local history. As part of this process of heritage promotion, the organizers gave out free key chains to heritage event participants (including those who visited JCCAC). I collected such souvenirs in my observation study. The Heritage Office's initiative was supported by local funding sources. Stated in the Hong Kong Jockey Club (HKJC)'s official website, the JCCAC was created with funding of HKD$94.4 million from HKJC that gentrified and retrofitted an empty Shek Kip Mei nine-floored industrial estate to incubate young and budding artists.[19] The conversion of an industrial building to a creative production center paralleled Hong Kong's transition from a light manufacturing sector (where industries have shifted to the Pearl River Delta region) to a service (finance and banking) and creative economy. With one of the highest GDP per capita incomes in the world and affluence as one of the four developed tiger or dragon economies, art consumption increased in Hong Kong and corresponded with its transition from manufacturing to service economy. Using Beijing's 798 (art district) as a case study, Jennifer Currier characterizes this transformation (industrial to consumption) as a "structural change from a production-based economy to a consumption-based economy," something that is equally applicable to all first-tiered cities in East Asia.[20] In the retrofitted building, modern features could be found in many areas but some of the more significant ones include a glass ceiling overlooking the building lobby with the penetration of sunlight lighting up the lobby floor where art sales are held.

OBSERVATION OF THE FACILITIES

By the time I visited JCCAC, I had already observed other creative villages in Hong Kong (Fo Tan and Chai Wan clusters of ceramicists in industrial parks). In Japan, I had also observed production communities in Arita and a mountainous pottery-making facility in Izumi, learnt about the Shigaraki artist-in-residency program from oral interview work and researched on the Sarayama folk potters community in the seminal work of Brian Moeran.[21] The methodology used in this paper is based on observation studies, textual collection and analysis and oral interview materials. The difference between JCCAC and other creative villages that I have learnt about can be summarized into three main points. The main difference between JCCAC and Sarayama, Arita, Shigaraki and Izumi case studies is that the latter four Japanese examples are specialized in pottery and ceramics arts production while JCCAC is a comprehensive multimedia production site with an array of art works not restricted exclusively to ceramics. Like JCCAC, situating artist clusters in Fo Tan and Chai Wan industrial parks probably originated from pragmatic reasons—space constraints and convenient clustering of artists for collaborative work.

Hong Kong's densely-populated environment results in high-density use of land and space. Therefore, all kinds of artistic production are neatly apportioned into space-saving multi-story industrial units found in JCCAC, Fo Tan and Chai Wan. The clustering of photographers serves the needs of a number of media-based companies found in Chai Wan's industrial area but they have also found ways to work with other artists that require photographic services as well. My informant Ms. A collaborated with a well-known Hong Kong-based photographer to create her exhibition catalogue and such collaborations arose out of personal friendships and networking. The collaborative opportunities and comparatively larger space of industrial units compared with city (especially central business district) retail units account for the reasons why art galleries keen on expansion are moving into the Chai Wan area. When I attended the opening show of an informant (friend of Ms. A) belonging to the fourth generation of ceramicists in Happy Valley, I learnt from the gallery owner that he had already decided to expand into Chai Wan. It seemed to be a market-driven choice, given that the Happy Valley gallery rentals are at a premium and spatial utilization has reached maturity in centralized city areas.

The array of art-forms found in JCCAC is probably most diversified amongst the art clusters covered in the paper. I was able to conclude this from the wide array of textual materials collected for the project. For the textual analysis section of the paper, I collected textual materials from JCCAC that reflected the eclectic nature of the fine arts community in JCCAC. Fine arts-related materials collected from my observation trips included topics related to culinary arts, tea arts, postcards on dance performances, photographic exhibition, textile designing, short film production and wine tasting, barista training, all sorts of drawing media, calligraphy, etc. in addition to pottery-making culture. The programs are also designed for all age groups, including children who join pottery classes. In other words, there is something for everyone: families with kids, traditional arts enthusiasts, cutting-edge experimental budding artists and yuppies who desire avant garde art consumption. In one such workshop on pottery-making for kids, a participant's comment included in its brochure noted that the students were able to learn about the lifestyles of pre-modern societies in the history of China as well as Jingdezhen blue and white ceramic cultures through the pottery lessons.[22] Because of the variety of art mediums and art courses found in the building, equipment supply is equally important so that artists are able to obtain the sort of tools and ingredients they need for artistic production. Equipment makers also provide information on equipment and materials supplies. The JCCAC is also an information resource, distributing free reading materials such as the *Gourmet Magazine* which featured the latest information on the culinary scene in Hong Kong, local and foreign dishes, cooking equipment and gourmet food supplies.[23] I also attended and observed a Chinese tea ceremony using Jingdezhen ceramics to brew fine Chinese tea. The brochures that I collected at the tea ceremonial site reflected cosmopolitan tastes when it comes to tea-drinking with printed information on Japanese, British and Indian teas in addition to the Chinese offerings. During the tea ceremony, I observed how ceramic arts (material culture) are combined with traditional brewing techniques (performance art) to make tea with artistic and aesthetic overtones. The infusion process highlighted the importance of ceramic designs. In the Chinese tea ceremony demonstration, the tea master poured hot water into a justice cup to evenly distribute tea concentration before serving. The Jingdezhen porcelain tray with hand-painted *sanshui* (*sansui* in Japanese) drawing is used to hold the tea sets while the *zisha* (purple clay) teapots are used as main infusing bodies for tea brewing. The picture indicates the importance of highly-specialized tools used for every step in tea-brewing found in the ceremony. In terms of arts consumption, the specialized artistic wares are affordable enough for the averge tea arts enthusiasts (the sets cost anywhere between HKD60 to HKD600).

Second, both Fo Tan and Chai Wan artists communities are full-fledged industrial parks so light flatted industries, seafood distributors and logistics units co-exist with artists' studios. In this sense, because artist tenants are buffer fill-ins at times when occupancy is low, so they come in later at a time when industrial tenants were not forthcoming. For some observers, the late arrivals create a challenge as the artists occupy units that are vacated or not taken up by industrial tenants. Consequently, Chiu argues that these galleries and workshops are dispersed throughout an industrial estate which may not facilitate exchanges between members of that creative community.[24] In contrast, JCCAC is exclusively occupied by artists in different fields and an art school. In that way, JCCAC's software infrastructure of talents is attuned towards the specialized needs of artists. Third, Arita, Sarayama and Shigaraki are equally specialized creative production ecologies but JCCAC has a space-saving feature as a multi-storied flatted creative village converted from a former industrial estate. Arita, Sarayama and Shigaraki also have flatter horizontal landscapes (not structured into a single multi-story building) that integrate activities related to raw materials storage, processing, supplier and extraction facilities. JCCAC's tenants are mostly engaged in the design and production stages of the creative process with little pre-production activities like raw materials extraction and storage. In contrast, the potters and ceramicists in Sarayama and Arita source their own timber and prepare them for storage and then cut them into suitably-sized units for the firing process. The clay used for the process is also crushed at these sites.

The third feature lies in the similarity between JCCAC with its Arita, Izumi and Shigaraki Japanese counterparts in that all of them produce traditional and interpretive works (re-interpretation of traditional artworks by incorporating modern features). When I visited JCCAC, I observed how traditional craftsmen are applying lacquer onto traditional Chinese folding screens while neighboring units next door are creating postmodern artistic interpretative works. The fusion of traditional wares with innovative modern works is visible in Arita as the craftsmen produce perennial favorites, blue-and-white (*sometsuke*) traditional wares, alongside modern interpretations of those items. Similarly, according to my informants, Shigaraki encourage artists-in-residency to create modern art and, amongst the techniques used, traditional ash-accumulated glazes remain highly popular. The Izumi craftsman used a recreated *anagama* (climbing kiln) based on millennia-old designs to fire traditional Satsuma-yaki (wares) while innovating new shapes like Western tea-cups amongst other non-East Asian shapes and forms. The combination of East-West and pre-modern/modern hybridized works in these production ecologies encourage constant innovation in traditional forms. Traditional works are constantly re-interpreted for a contemporary audience and consumer market. The constant innovation ensures the continued relevance of crafts made by the artists for a mass consumption society with options to acquire cheaper utilitarian wares available for home delivery through online sales or retail sales from shops that import them. In an age of mass consumption, affordable but high quality and unique pieces are likely to find consumer interest from middle income classes of consumers. One of my informants opined that artistic concepts with well thought-out designs are the main difference between wares made in these artist clusters and mass-produced items manufactured in industrial estates.

ENVIRONMENTALLY-FRIENDLY FEATURES

Like Chai Wan, JCCAC is surrounded by residential areas and slopes but not bereft of greenery. In the case of Chai Wan, I was able to spot hillside trees when I glanced out of the window of my informant Ms. A's studio. According to Chiu, communal green spaces were integrated into the JCCAC structure.[25] And, in response to feedback from the neighborhood, JCCAC cut down the number of publicly-visible lighting in the building and only uses the bare minimum lights in the after-dark hours (this incidentally saved energy generation bills by approximately 20%).[26] In terms of green corners in the building, the tenants try to cultivate plant-life in various green areas of the building whenever possible. In some areas, rows of green plants line the corridor walls. They are placed in artistically-selected dark green pots.

Environmentally-friendly initiatives can also be found in artistic works themselves. When the JCCAC project was first conceived, a green initiative in the facility's revitalization process was to recycle and reuse undistributed and unsold manufactured items from the former industrial estate such as clothes hangers, timepiece components, dim sum steamers and candle-stands and turn them into fine art crafts and creative display items.[27] Some of these hand-made products can be found in the retail display shops on the lobby floor. In one workshop, participants learn how to use natural materials like ostrich egg for painting. Amongst the textual materials that I collected in the JCCAC Festival 2012, some of them featured works that were created using plastic trash and expired food. Therefore, there is conscious awareness of the importance of the environment amongst some artists who utilize the idea of the environment as the themes for their exhibitions or the conceptual foundation for their works. There are other works that relies on making contact with the Hong Kong agricultural community, tapping into their expertise to produce video images, and cultivating indoor agriculture in JCCAC (serving as an example of urban gardening for other densely-built urban environments in Hong Kong).

OBSERVATION STUDIES – A SIDE GALLERY CASE STUDY

The networking and social events that I attended were cordial and interactive. In the first show that I attended organized by Ms. A, the works were displayed at a side gallery which was space-saving in high-rent Hong Kong. Stated in the mailer, the work was surrealistic and combined scientific rationality and artistic pulses to spur audiences to ponder about the state of the contemporary universe and globalized world.[28] Part of the work on display reminded me of her 2011 mixed media installation of the LvLn virus series (probably where the inspiration for the ceramic heads in the current exhibition was derived from).[29] Before going to this exhibition, I was already exposed to Ms. B's works, particularly the "love virus" series (LvLn). The white clay in that series used to make the porcelain heads comes from Japan can also be found in this side gallery display. The lifeless scattered heads in the LvLn virus series has a contradictory apocalyptic imagery juxtaposed with a calm Zen-like serenity especially when it is displayed against the background music of a Gregorian chant (featured in the official video that the artist shared with me). Out of the scattered heads installation, a lonely tree emerges with branches that integrate embedded hands reaching out in an equally Zen-like, surreal, apocalyptic and calm manner. The ambiguity of the images conveyed by the display evokes mixed emotions in the audience as they try to understand how viruses represented by suspended ceramic heads with aerodynamic designs infect, cross-pollinate, penetrate, replicate, self-assemble themselves. The branches of life, the suspended humanoid ceramic heads and out-reached hands and limbs that exude water droplets all carry connotations of the complexities of nature and life engaged in biological processes expressed in a poetic manner.

On December 9, 2012, when I went for the opening show, I interacted with enthusiastic fans and supporters of Ms. A's works. The crowds that gathered outside the side gallery indicated tremendous public artistic interest in Ms. A's show. On December 15, 2012, I returned to the side gallery so that I can experience the same display without the crowd and in solitude. In the absence of Ms. A or her supporters, when I took photos of the display, I overheard passers-by (visitors who came to JCCAC to view the arts festival) commenting on the piece, indicating the complexity of the mixed media work and how it was conceptually attractive and visually stunning. They also speculated on the type of materials used to make the various components of the side gallery display (the "Silence" series for example). These passers-by resembled young budding artists out and about in a cultural village to obtain inspirations from their established counterparts. In the multidisciplinary, multimedia and multi-faceted fine arts atmosphere of JCCAC, it is unlikely that Ms. A will face the kind of "pseudo-ceramics" critiques which were mentioned by her former mentor in the catalogue of the LvLn series of work.[30] In fact, in line with my observation on December 15, 2012, Ms. A is likely to find new fans for her works at the facility.

OBSERVATION STUDIES – WORKS CREATED IN SHIGARAKI JAPAN

In the second gallery event that I attended, the latest works of ceramicist Ms. B created during her time in Japan (Shigaraki) was displayed at gallery on December 15, 2012. The subtle Zen-like features of lotuses, paper screens and curtains made out of delicately-thin ceramics conveyed a serene and peaceful sense of aesthetics. The serenity was enhanced by the background music compiled from everyday sounds recorded by Ms. B while she was an artist-in-residence in Shigaraki. The lifelike lotus made out of thinly-crafted ceramic films added a touch of rustic nature to the tranquil display of earth-tone colored ceramics. Ms. B revealed to me that the copper stands (representing stem and base) holding the lotus cup were specially made with high-quality copper (more expensive than those found in Hong Kong) crafted by an experienced artisan in Shigaraki for Ms. B. Ms. B was grateful and postal mailed her exhibition materials to him as a sign of respect for his contributions and help to the project. During my visit to Ms. B's opening show, I made intimate visual contact with the lotus cup art piece and was particularly struck by the lifelike pods in the lotus. One of the lotus cups had the "palladium silver effect" represented by a metallic covering over the surface of the lotus cup which gave one of the two lotus cups a lustrous shine. Contrasting the heavy lotus cup pottery with another work entitled "curtain" (a porcelain representation of paper screens used in pre-modern Northeast Asia), one could detect Ms. B's skills versatility in crafting items ranging from bulky and heavy ceramic pieces to delicately thin porcelain works where light shone through easily. The *shoji* paper screen ceramic piece with its unique texture captures the complexities of light penetration through actual paper screens. The intermediation of the paper medium creates a mellow and warm glow after light passes through the screen. Paper screens retain warmth emanating from the light source while casting a warm glow when the light shines through the screen. Ms. B's piece captures this mediated light source and the feeling of warmth.

LEARNING EXPERIENCES IN SMALL-SCALE ARTS ENTREPRENEURSHIP

From the textual materials I collected on December 9, 2012, I was able to spot the participation of educational as well as non-profit organizations in JCCAC's academic activities. There are schools in the facilities that offer courses related to arts and I also managed to collect brochures and pamphlets related to continuing education as well as diploma courses. In this aspect, JCCAC also serves as an information center in arts education. In the non-profit sector, entities include the Hong Kong Society for Education in Art which is based on the second floor of the building and it has an eclectic agenda in education that advocate quality improvements in art education; upgrading public recognition for the industry's workers; hosting art-related events; and adopting a global perspective of fine arts advocacy.

Cultural lessons aside, the facility also organizes a handicraft fair which hones fine arts entrepreneurial skills for budding artists and artisans. I attended and observed this fair known as the JCCAC Festival 2012 on December 15, 2012. The event was publicized in a free art magazine Artplus (circulation: 40,000) and Artmap distributed free throughout Hong Kong and Macau in addition to JCCAC's own brochures and handouts.[31] During my observation session, the main lobby on the ground floor was occupied mostly by jewelry, porcelain, glass works and calligraphic works of art on display. In the enclosed gallery spaces by the side of the open lobby grounds, the bazaar participants specialized in handmade leather products. The highly-specialized displays zoned according to the categories of works on sale had the advantage of clustering similar types of artistic works for the convenience of viewing by public participants of the fair. The artists distributed name cards, flyers, posters, postcards and also printed materials with social media contact details printed on them. The bazaar was fairly international as I conversed with artists from China and European locations. The foreign artists mentioned to me that their products were fired or made in the JCCAC facility itself, indicating the presence of kilns in the facility. A variety of techniques were used by the bazaar's participants in making their products including carvings incised on the surface of the fired clay, application of cobalt for the blue and white effect and application of wax for creating pottery designs. The event had a festive feel as members of the public crowd enthusiastically around artistic works, asking the artists questions about their pieces, acquiring new decorative pieces for their use - the net effect is all participants eagerly getting exposure to arts and culture. The organizers also distributed specially-designed stickers with the event emblem to participants as souvenir keepsakes. On the same day as the bazaar, the studios in JCCAC opened their doors to members of the public for viewing.

I presented this chapter at the New York Conference on Asian Studies (NYCAS) in State University of New York (SUNY) Binghamton on September 28, 2013 and obtained important feedback. The first question is related to the funding of JCCAC. The poser of the question wanted to know if the funding for JCCAC comes from the state. According to the official website, government funding inputs are minimal as the funding comes mostly from a charitable organization, The Hong Kong Jockey Club Charities Trust (JCCT), rather than state sources.[32] In terms of management, it is run by a university rather than a state agency. The second question was whether a state-driven or non-state-driven initiative works better for JCCAC. Drawing from my main text, my response was that it takes close collaboration between the two through consultation to draw up the optimal policies for making JCCAC a success. The third question that I received was how much governmental intervention is acceptable for the JCCAC community. I provided the example of the dimming down of bright lights as an example of sensitivity to local community needs. Governmental intervention alone would not have convinced the artists to autonomously and independently decide on this initiative. It took a tripartite decision making process involving the artistic community, state and local resident feedback mechanisms to achieve an optimal solution that is a win-win situation for the community. Overall, a tripartite consultative mechanism appears to be the working solution to the JCCAC community's challenges. The success of JCCAC would not be possible without cheaper rents made possible by state policies. For the state, gentrification of an old industrial building located amidst an aging housing estate is also beneficial to urban planning for older areas of the city. Its output cannot be fully realized without the artistic freedom and space for the art creators. JCCAC also needs the understanding of local community to function without resistance from nearby residents. In other words, all stakeholders need to reach some form of equilibrium in this arrangement.

NOTES

1. This chapter is funded by the South China Programme Research Grant.

2. Housing Authority, "Historical Background of Shek Kip Mei Estate," accessed December 12, 2012, http://www.housingauthority.gov.hk/hdw/en/aboutus/events/community/heritage/about.html.

3. Ibid.

4. John McCarthy, "Regeneration of Cultural Quarters: Public Art for Place Image or Place Identity?" *Journal of Urban Design* 11, no. 2 (2006): 243; Graeme Evans, "Measure for Measure: Evaluating the Evidence of Culture's Contribution to Regeneration," *Urban Studies*, 42, no. 5–6 (2005): 959–83.

5. Ibid.

6. Ping Kwan Chiu, "A Unique Experience of Setting Up and Running the HK First Artist Village" (Powerpoint presentation at the Half Day Symposium on Revitalisation of Factory and Old Buildings, Hong Kong Institute of Facility Management, January 22, 2010), 6, accessed December 13, 2012, http://www.hkifm.org.hk/2010_doc/ppt_20100122_pkchiu.pdf.

7. David Ley, "Artists, Aestheticisation and the Field of Gentrification," *Urban Studies* 40, no. 12 (2003): 2530.

8. Walter Santagata, "Cultural Districts, Property Rights and Sustainable Economic Growth," *International Journal of Urban and Regional Research* 26, no. 1 (2002): 9.

9. See Chiu, "A Unique Experience of Setting Up and Running the HK First Artist Village."

10. Desmond Hui, "From Cultural to Creative Industries Strategies for Chaoyang District, Beijing," *International Journal of Cultural Studies* 9, no. 3 (2006): 318.

11. Information Services Department, Hong Kong Special Administrative Region Government, "Creative Industries," accessed December 17, 2012, http://www.gov.hk/en/about/abouthk/factsheets/docs/creative_industries.pdf.

12. Jason Hackworth, and Niel Smith, "The Changing State of Gentrification," *Tijdschrift voor Economische en Sociale Geografie* 92, no. 4 (2001): 464-77; Hui, "From Cultural to Creative Industries Strategies for Chaoyang District, Beijing"; Jennifer Currier, "Art and Power in the New China: An Exploration of Beijing's 798 District and Its Implications for Contemporary Urbanism," *Town Planning Review* 79, no. 2–3 (2008): 237–65; Santagata, "Cultural Districts, Property Rights and Sustainable Economic Growth"; Valerie Kinloch, "Youth Representations of Community, Art, and Struggle in Harlem," *New Directions for Adult and Continuing Education* 116 (2007): 37–49.

13. Chiu, "A Unique Experience of Setting Up and Running the HK First Artist Village," 14.

14. Luciana Lazzaretti, "City of Art as a High Culture Local System and Cultural Districtualization Processes: The Cluster of Art Restoration in Florence," *International Journal of Urban and Regional Research* 27, no. 3 (2003): 635–48.

15. Evans, "Measure for Measure."

16. Chiu, "A Unique Experience of Setting Up and Running the HK First Artist Village," 1, 9.

17. Ibid.; Hulu Culture, "JCCAC," accessed December 13, 2012, http://www.huluhk.org/SSP/heritage.php?id=jccac&lang=en.

18. Commissioner for Heritage's Office, Development Bureau, *Heritage Fiesta 2012* (Hong Kong: Commissioner for Heritage's Office Development Bureau, 2012).

19. Hong Kong Jockey Club (HKJC), "Explore 'Old Factory, Emerging Arts' Exhibition at HK Racing Museum," December 16, 2008, accessed December 13, 2012, http://corporate.hkjc.com/corporate/corporate-news/english/2008-12/news_2008121620748.aspx.

20. Currier, "Art and Power in the New China," 237.

21. Moeran, *Folk Art Potters of Japan*.

22. Hang Seng and Jockey Club Creative Arts Centre, *Hang Seng Arts Empowerment Workshop Exhibition* (Hong Kong: Hang Seng and Jockey club Creative Arts Centre, 2012), 28.

23. Deli Communications, *Gourmet Issue* 40 and 41 (Hong Kong: Deli Communications, October and November 2012).

24. Chiu, "A Unique Experience of Setting Up and Running the HK First Artist Village," 6.

25. Ibid., 23.

26. Ibid., 24.

27. HKJC, "Explore 'Old Factory, Emerging Arts' Exhibition."

28. Wai Wai Yim, *Front Side Gallery Invitation* (Yim Wai Wai's Solo Exhibition pamphlet) (Hong Kong, December 9, 2012).

29. Wai Wai Yim, " 'LvLn Virus' Ceramics Installation by Yim Wai Wai," accessed December 13, 2012, http://www.youtube.com/watch?v=1oZjkrUMmIA.

30. Wai Wai Yim, *LvLn V irus* (Hong Kong: iSquare Design and Development Ltd, 2011), 18.

31. Artmap, *2012 Artmap* (Hong Kong: Artmap, 2012).

32. Hong Kong Baptist University (HKBU), "Jockey Club Creative Arts Centre," accessed October 15, 2013, http://gao.hkbu.edu.hk/en/sps/jccac.html.

Chapter Nine

Fieldwork in Arita

The Ecology of the Town, Its People, the Natural Environment, and Its Energy Use

This chapter is based on my fieldwork carried out in the towns of Arita and Imari Japan in the summer of 2013.[1] It is the first case study of a specialized pottery-making town in the publication. My intention is to write an empirically-rich chapter that records the observation studies and unstructured open interviews with the residents of Arita and Imari. It will discuss some of the important points brought up in the earlier chapters. The ecology of stakeholders that I covered and interacted with during my fieldwork included the ceramics artists, the large-scale ceramics manufacturers, the consumers, the ceramics researchers, etc. I interacted with individuals and institutions and observed the contributions they made to Arita's (and Imari's) ceramics production. Speaking to the stakeholders engaged in this industry, there are two broad categorizations of ceramics production in Arita – the artistic production sector and the utilitarian wares mass manufacturing sector. In other words, there was creative space and production capacity in the town to accommodate both small-time, individual or even amateurish potters as well as world-class large-scale ceramics manufacturers. This is a technique that I derived from social history. To understand the ecology of a community or society, it is necessary to interact with both the elites as well as the non-elites of various economies of scale, social class and the marginalized.

To understand the products made by Arita and Imari residents, I sought to understand how they defined their own products. This is a complicated area to navigate given that both towns have a complex combination of small-scale artistic producers and world-class ceramic makers on an industrial scale. Interestingly, throughout my trip, there was a conscious effort in the narratives by stakeholders in the Arita community to distinguish their products from *mingei* (folk arts). Many of them opined that folk art ceramics was considered crude, made for rough use and not mass-produced. Some of my informants defined Arita products as mass-produced utilitarian wares with a separate artistic category for individual works. A prominent practitioner of the town of Arita defined its products in the following way. Arita ceramics or "Arita-yaki" is defined in a textual material in Master Inoue-Manji's (a "Living National Treasure" residing in Arita) Hong Kong Sogo exhibition in early December 2012 as porcelain made in the town of Arita and its periphery of Nishimatsuura district of Saga Prefecture.

While there are some ongoing debates about how Arita products are defined, almost all the stakeholders that I met with agreed that Arita was an early site for Japan's early industrialization. Just as Arita had been exporting utilitarian kitchen wares to the global consumers for the past few centuries, I arrived at a time when Arita was gearing up for greater global reach for its products. The producers in Arita were interested in reaching out to the consumers in the global market. They enjoy one historical advantage: Arita or Imari-style[2] porcelains have been exported to the rest of the world since the sixteenth century. A testimony to this is the impressive display of export wares in the Kyushu Ceramic Museum (including chandelier, vases, plates, etc.) as well as the palaces of Europe (the Topkapi palace in Turkey, as well as Dutch, Portugal and German palaces).[3] Visitors interested in viewing such export wares can also pay a visit to the full-scale mock-up of Zwinger palace in the Arita porcelain park where some pieces are on public display.

The initial institution that facilitated Arita's globalization was the trading ships of the VOC (Dutch East Indies Company) which carried Imari export wares to Europe starting from the mid-seventeenth century and bypassing parts of Southeast Asia and India in the process of conducting entrepot trade. In the early modern period (1868 to the early twentieth century), a major promoter of Arita porcelains to the West were the international fairs organized in cities like Paris, Philadelphia, etc. In these fairs, Arita products won international acclaim and recognition by consumers in the West. According to written texts displayed in Master Inoue Manji's exhibition in Hong Kong, Arita products exhibited by the late Tokugawa Saga Clan (also known as Nabeshima-han) attained the Gold Medallion and Grand Prize held in Vienna and Philadelphia. The combined efforts of maritime trade and exposure in international expositions effectively globalized Arita products in the Western modern world (contemporaneous to Japan's late pre-modern or early modern period). Given the important contributions that the Europeans, especially the Dutch, made to the exportation of Arita/Imari wares, the imagery of the Dutch, their ships and traders coming to Japan remain firmly etched in the consciousness of some classical ceramics designs even today. Arita/Imari export wares to Europe are displayed in many fine museums and galleries found in the towns of Arita and Imari.

While export wares excited the ordinary consumers, the marketing of Koransha's showroom pieces at international expositions drew the attention of large-scale traders, collectors and trading companies. The expositions provided the necessary exposure for Arita/Imari pieces for a Western audience, particularly the commercial sector. The atmosphere, products and display from this era are reproduced in the showroom galleries in Fukagawa's China on the Park as well as other show windows.

PHYSICAL FEATURES

In terms of physical features, as one drives or gets into Arita by car or train, the sights that first greet the visitor are the rice fields as well as mountains and hills. In fact, the former President of Koransha Fukagawa Tadashi described Arita as a "gorge town (*sankyō no machi*)" in his publication *Koimari no bi to roman* [Old-Imari's beauty and romance],[4] a town located at a narrow corridor in between steep rocky slopes of a ravine which is surrounded by mountains with the residents habiting a low lying area. Low-lying two-storied houses dot both Arita and Imari, some have solar panels installed on the roofs (probably used to power hot water boilers for showers, lighting and heaters), creating a serene small-town or city atmosphere conducive for both creative production and tourism. Some of the low-lying houses have kiln chimneys, in many cases, a reminder of the past when timber and coal were used to fire ceramics. In both cases, gas kilns that have invisible and environmentally-friendly emissions are used. The chimneys are no longer in use. The iconic Arita Koransha company as well as the historic Imari Seizan (currently managed by its fifth generation descendants) pottery studio that I visited both had such red-bricked chimneys but they are no longer functional and mainly retained as nostalgic artifacts.

In terms of class divisions, there is a sense of egalitarianism in the Arita community as there are no particularly visible signs of outstanding and ostentatious lifestyles amongst residents. Even the external facade of Koransha's Fukugawa family home found inside its headquarters compound is subtle, tasteful, austere and well-hidden in Koransha's headquarters without the trappings of opulent wealth. Many residences have also installed solar panels on the roof to conserve energy.

Geographically, the physical location of contemporary Arita is landlocked while the town of Imari has access to the Bay of Imari which allows the town to export ceramic wares to other parts of Japan in the past. In Master Inoue Manji's exhibition in Hong Kong, an exhibition signage indicated that the term "Imari" was defined as porcelains that were shipped from the port of Imari and made in the porcelain industry of Arita. The Seizan pottery studio once historically served the porcelain supply needs of the Nabeshima-han. This particular class of porcelains was made only for *daimyos* (the feudal lords) in the Nabeshima region. They are not accessible to commoners. In rare cases, Arita wares with attractive designs were hand-picked for elevation to Nabeshima-ware status and they cease to become accessible for consumption by commoners after that. Patronage by the Nabeshima-han (also known as Saga Clan or Nabeshima clan) was important in this way since they gave certain wares a special status that distinguished them from other ordinary Arita wares. The Fukagawa family is a good example of a beneficiary of patronage by the Saga Clan in the Tokugawa period, although their fortunes changed when the Han system was abolished to make way for the Emperor system in the early modern Meiji period. The Fukagawa family was skillful enough to switch their patron to the imperial household, a relationship that remains intact till this day. The town of Imari had similarities and differences from Arita. Both have traditional sights but Imari preserved the heritage found in a mountainous hamlet around the iconic *Seizan* studio in a location that was designated ceramics potter for the daimyo who used to rule the area in the late Tokugawa period (now part of the Okawachiyama region which is known for its natural beauty). Proximate to each other, both Arita and Imari towns are beautifully lined with rivers, mountainous landscapes and green spaces, particularly in the suburban areas away from downtown. The natural environment and a panoramic sight of nearby mountains was a secondary consideration for Koransha to build their current Heisei-era factory on a hill in Arita.

ECOLOGY OF PRODUCTION

According to my informants at the Folklore Museum of Arita, rather than one single hegemonic communal group-oriented community, Arita functions more like a collection of factional groups, each with their own projects and interests. Unlike other pottery-making towns, Arita is not organized like a hamlet with close inter-linking relations between individuals but in terms of individual producers or along project work groups. When I visited Arita's specialized folklore museum (The Folklore Museum of Arita) and posed the question of the existence of a tightly-knit network of potters in Arita, my informants responded by saying that the community is organized according to interest-driven group projects rather than tightly-knit communitarian hamlets. Rivalries were also not highly visible, except through sale of products and, even so, such competitions were based on market forces. Ordinary consumers and collectors tend to compare the product lines of Fukagawa and

Koransha for example, but this does not imply rivalry between these two great porcelain makers.

Overall, it can be characterized as friendly competition based on market forces. Both firms originated from the same Fukagawa family line although somewhere along the lineage, two brothers decided to set up different kilns that resulted in the separate entities of Fukagawa and Koransha today. The split was, however, completely amicable as I understood from my informants at both porcelain makers' facilities during my fieldwork. Aside from competing for consumer market share, cooperation between ceramicists and porcelain firms were not uncommon, given the collection of talents and expertise in the towns of Arita and Imari. An incentive for cooperation is the size and scale of a given project. The bigger the project, the greater the need for mobilizing collective efforts and resources, especially when its needs go beyond the capacity of a single individual talent, regardless of that individual's genius-ness.

Besides the producers, there are other specialists in the Arita and Imari locations who do not produce tangible outputs but are equally important in preserving older but inspiring ideas for future generations of potters and ceramicists. The archeologists, collectors, historians and scholars are the recorders in this ecology of production, collecting shards, interpreting designs and recording community life to preserve the stories behind the production processes in Arita. Curators, for example, keep samples of pre-modern Imari and Arita pieces intact for future study. A number of my informants who are curators and folklore researchers are trained in the field of archeologists. The Yuko and Akihiko Shibata Collection in the Kyushu Ceramic Museum is probably the biggest collection of Arita and Imari specialized in Edo to Bakumatsu porcelain pieces in the world (with 10,311 pieces) and it has become a nationally-recognized tangible cultural asset since March 18, 2006. Besides state or prefecture-managed facilities, there are private galleries and collections that are equally historical and exquisite. One example is Koransha's gallery showroom located at its headquarters.

Efforts were made by practitioners in both Arita and Imari communities to rebuild *noborigama* or climbing kilns to preserve the traditional art of firing ceramics using timber. Communitarian activities like these bonded the residents of the Arita and Imari communities respectively. The biggest communitarian event is probably the annual sale in Arita where, once a year, visitors numbering approximately one million (according to my informants from the Saga Prefectural government), thronged the streets of downtown Arita during the famous porcelain sale. According to a *Japan Times* article, organized by the Arita Chamber of Commerce and Industry, the yearly Arita Ceramic Bazaar in Arita can reach 3 billion yen in sales.[5] My informants told me that consumption during this special period amounted to more than one year's worth of sales outside the annual fair. Therefore, there is both economic incentive and social bonding factors to unite the town and gear up for the visitors streaming in from other parts of Japan and from overseas.

The sales and distribution network in Arita was extremely important. Public exposure accounted for the difference between brand names that were successful and those that made less sales. For example, larger firms in Arita like Fukagawa have vans that carry pictures of ceramics and their company logo, serving as mobile advertising platforms for the company and its products. Sometimes, the concept of *gentei* or limited editions distinguished wares made especially for and sold only in the headquarters (*honten*) of well-known bone chinaware makers from those sold in other shops or branches. This encourages visitors and tourists to consume these *gentei* products as keepsakes from their visits to Arita. Besides *gentei*, the idea of *natsukashī* (nostalgia) is another marketing tool for sales and distribution.

Research resources on ceramics are also available in the production ecology of Arita. Three examples are the social scientific institution of The Folklore Museum of Arita; the educational institution of the Arita College of Ceramics; and the R&D (Research and Development)-based Ceramics Research Institute. The Folklore Museum of Arita retains and records the use of ceramics in the daily lives of Arita residents in the town's history. It was here that I gathered historical materials of bygone technologies like clay crushers, understood pre-modern methods of ceramics production (*noborigama* operations) and by making visual contact with the equipment and mines. I was able to comprehend early postwar development of the community through my interviews and conversations with informants.

The second example is the Arita College of Ceramics which has an inbuilt library resource and a collection that stores and displays ceramics from all parts of the world. Besides a community-based function of imparting skills from experienced artisans to willing students, the College serves as a platform for foreign students to understudy Arita methodologies in ceramics-making.

The Arita ecology also included the technologists. Some of them work in the Ceramics Research Institute. This is the scientific research nucleus of the town, producing the latest ceramics products and other high-tech applications. The center is connected with other establishments in different ways. For example, it has the technological capability to examine ceramic shards to determine their age. They also have specialized machineries to produce cutting-edge ceramics products. The Ceramics Research Institute and the Arita College of Ceramics are discussed in greater details below.

TECHNOLOGY

In my fieldwork, I enquired if my informants recognized the fact that modernization in Arita was attributable to Western engineers like German chemist and engineer Gottfried Wagner and they replied in the affirmative. My informant at the Kyushu Ceramic Museum noted that Wagner was a chemist and one of the technologies that he introduced to Arita was the blue ink that is used to fire blue and white porcelain. Previously, the cobalt needed for the blue color found on fired porcelain had to be imported, just as China did, from Persia. But Wagner's transfer of knowledge made it possible for cobalt to be produced in the laboratory.

Because of Western knowledge and technologies, mass production (*tairyōseisan*) of ceramics became possible. It also phased out pre-modern modes of production. Some of the pre-modern or modern technologies became obsolete in the modern period, for example, the waterwheels that are used to crush the clay. The crushers are no longer a source of noise pollution. In fact, the only water-powered clay crusher that I saw in Arita was located outside the Folklore Museum of Arita, now a heritage artifact. Another reason for crushers, electric or water-powered, to fall into disuse in the contemporary period is due to the fact that even the clay used in Arita as well as the studios that I visited in Imari no longer use Arita clay. This came as a surprise to me given that the historical origin of the town itself was based on the discovery of the clay. Arita was founded when kaolinite was discovered by Ri Sam Pei in Izumiyama at the start of the seventeenth century and the hill has since been designated as "national historic landmark" in 1980.

Arita's clay supply now comes from Amakusa in Kumamoto.[6] The reason for this is mainly due to the better-quality Amakusa clay that did not require wood dust mixture for kneading the clay which Arita clay needed. Another school of thought that I encountered in my fieldwork is that Izumiyama porcelain stones have been depleted. The plaque display at the Izumiyama site mentions the saying that the entire clay mountain of Izumiyama has been turned into ceramics! According to the plaque installed at the Izumiyama site, even though the site potentially contains kaolinite,[7] the mine is mothballed because of challenges in managing waste water disposal and drainage. The doomed fate of the mine therefore rests on three perspectives: the mining industries' difficulties in water resource management; the ceramicists' perspectives of the quality of Izumiyama rocks; and my informants' hearsay evidence of mine depletion.

Older technologies have also disappeared from the landscape. I searched hard for the traditional *noborigama* or climbing kilns in Imari that once dotted the region but was unable to locate them. Early modern industrialized workplace has been transformed into a high-tech factory system. The clay crushers powered by water are now museum artifacts and replaced by clay imported from Amakusa. The caption (made of porcelain) for the crusher on display outside The Folklore Museum of Arita indicated that the crusher is capable of using waterpower to crush 120-180 kg worth of porcelain stones each occasion. Over 100 crushers were located in Izumiyama, Shirakawa, etc. They were later replaced by electrical-powered crushers.

The same division of labor still exists and the specialization of tasks remains relevant in the contemporary factory system but there are significant differences. First, the use of technologies has made the system more efficient, streamlined the workforce and facilitated multi-tasking. Laser pointers, for example, helped workers to accurately draw and sketch designs before they are painted with iron oxide for firing. From my fieldwork observation, there appears to be little resistance towards greater technological adoption. It is easier for Arita craftsmen to accept the importance of mechanization, rationalization and productivity because Arita has a long history of mass production in ceramics, unlike other ceramics-making clusters specialized in artistic production where manual inputs, the human feel and touch are still preferred over mechanization. In other words, these are locations where aesthetics is prioritized over functionality and utilitarianism. Because of Arita's long historical experience in producing export wares, such notions were not held without qualifications. In addition, Arita and Imari wares do not carry the baggage of hegemonic historical traditions in the sense that they are very often classified as *kinsei* (early modern) pottery.[8] Therefore, the association of Arita and Imari with the concepts of modernity (early to late), mass production (proto-mechanized to mechanized), mechanization (water-powered to electrically powered) and factory system (late pre-modern to Taylorism) is historically and intimately intertwined compared with other forms of pottery.

While the dichotomy between primitive machines and advanced technologies is clear, recognition of the importance of heritage, culture and history prompted both Arita and Imari to retain their timber-fired *noborigama* (climbing kiln). One surviving climbing kiln in Arita is managed by Fukagawa and operated once a year because the management wanted all staff members to experience the unique and laborious timber-firing process as a form of cultural awareness. The products made from this timber-firing are sold at Fukagawa's gallery at China on the Park but at much higher prices than the normal gas-fired product lines because of the intense efforts needed for timber-firing and also the higher quantity of defects. I visited a climbing kiln in Imari which is operated for the same reasons. The community used the climbing kiln for firing several times a year to retain the skills needed. The cooperative kiln is also communally-managed so the entire community is involved in its upkeep. In both cases, there is a clear sense of camaraderie and community bonding.

While useful for firing large quantities of ceramics, the *noborigama* (climbing) kilns were still fired using timber and did not have precise controls. Precise temperatures were important for Arita production since Arita is well known for its porcelain wares (or *jiki*) which require the highest temperatures amongst pottery wares for firing. Technology for hi-firing made it possible for Arita ceramicists and potters to fire porcelain (*jiki*) between 1300-1400 degrees Celsius. When tapped, *jiki* gives a light crisp melodious sound. Porcelain is generally distinguished from earthenware (fired at 800 degrees Celsius) and stoneware (1000-1300 degrees Celsius). Advanced technologies facilitate the firing process by enabling greater precision in temperature control with less defects and better ergonomic controls (e.g., using digital technologies). High-quality fine bone china porcelains require additional care in the firing process. In the contemporary context, they involve two-staged firings: the preparatory firing (*suyaki*, about 900 degrees Celsius) and the main firing (*honyaki*, about 1300 degrees Celsius). In the case of over-glaze porcelains, it takes at least up to three firings. I was able to compare pre-modern methods of production with mass production at the high-tech Koransha factory. Despite the presence of advanced technologies used in production, certain traditional techniques and human touches continue to characterize Arita wares.

INNOVATION

Traditionally, the sources of influences for Arita designs came from continental East Asia[9] but, in the course of exporting to the West, it indigenized Western designs and, eventually, localized designs and unique patterns emerged. The Kyushu Ceramic Museum duly acknowledges the early historical influences on Arita coming from Korea and China. Continentally, Korean ceramics were amongst the first in the region to indigenize Chinese designs. After these ceramic influences spread to Japan from mainland Northeast Asia, by the middle of the 1600s, unique Arita designs began to emerge and these samples are viewable in the Shibata Collection housed in the Kyushu Ceramic Museum and private collections like the Koransha gallery.

Design preferences and sense of aesthetics are highly personal and subjective. I viewed the displays of artists designated as “living national treasures,” unknown amateur artists and also visited showrooms and sales galleries that display mass-produced utilitarian ware. While aesthetics is not a focus of this chapter, I will describe some of the new innovative Arita ceramics products and designs that I observed during my fieldwork. Innovation can be functional in nature, e.g., mass-produced rice bowls that are ultra-lightweight or can prevent rice from sticking to their bottom. Other innovations are closely related to health and hygiene when using porcelain wares that are either bacteria-resistant, made of safe recycled eco-friendly wastes or colored with organic dyes and safe chemicals which are harmless for consumption at the dinner table.

A variety of consumer ceramics products is displayed at the Ceramic Research Institute's showroom. Besides display pieces, the mass-produced items in Arita also extend to industrial products such as wall tiles, one of the array of ceramic products manufactured in the town. Innovation can also be design-driven. Some specialists like the Inoue Manji family concentrate on making *hakuji* (white wares) and created innovative designs based on traditional Japanese white ware designs. The eldest son of Inoue Manji mentioned to me that their designs are now based on incorporation of designs inspired by nature. There are only ten "Living National Treasures" nation-wide in Japan and Master Inoue Manji was given the title in 1995 in the area of white wares while Master Kakiemon passed away near the time of my fieldwork in Arita. My regret was not being able to pay a visit and interview the fourteenth generational namesake of the Kakiemon family "national living treasure" who passed away a few months before my 2013 fieldwork to Arita. When I visited the Arita College of Ceramic which the fourteenth generation Kakiemon successor headed as its principal, the college was the midst of a leadership transition to the fifteenth generation. Seniority and age were probably factors for consideration in this succession plan although this does not apply to all Arita and Imari[10] studios in the pottery-making communities. The eldest son of the Inoue Manji family, for example, is making a name for himself as the next-in-line namesake of the Inoue household. When I visited him, there was a queue of important official visitors waiting to see him next.

In this sense, for old established potter families, succession remains traditional and based on family/clan decisions, but there are no indications that all studios follow this format. In the case of the Inoue Manji (the current remaining "living national treasure" after the death of the fourteenth generational namesake of the Kakiemon family), his studio apprenticeship system is managed by his sons but also involved helpers and assistants who are not related to the family. Similarly, the Arita College of Ceramic accepts apprentices and students from all over Japan as well as foreign countries. When I was at the College, I saw the works of German and Korean students on display. In the Kyushu Ceramic Museum, I came across an exhibit that combined both functionality (clock function) with aesthetics (blue and white), a clock that could open up every half-hourly with rotating seasonally-displayed dolls placed inside.

Innovation could also be spotted in contemporary works of art. Sometimes, older techniques of firing and glazing ceramics are revived for producing contemporary pieces of art. One of the outstanding contemporary Arita work of art was designed with glazes colored using uranium compounds – an old technique that is revived in the contemporary period. The spirit of innovation is not only restricted to top artists and large firm or studios. Amateurs too have created designs that are functionally useful and aesthetically pleasing or creative. These works of art by amateur potters of students are visible for public display at the Arita College of Ceramics.

IDENTITY

Given the historical significance of Arita and Imari export ceramics products, it was natural to expect the nostalgic reproduction of popular evergreen designs. The idea of nostalgia or *natsukashii* quality was not as strong as I expected in terms of the creative thinking process of the contemporary producers that I had conversations with. However, physically, the quality was clearly embedded in the identity of the towns that I visited. For example, in the town of Imari, ceramics with pre-modern Imari designs could be found on display in public spaces, including hotels and other commercial spaces. Public spaces such as vending machines corners, train stations and museums are consciously designed with Arita ceramics motifs as part of the town's identity. In the town of Arita, restaurant kitchen wares, door handles, tiles, pads for electrical switches and other fixtures were often made of porcelain fired in Arita.

At the train station, showcases with glass displays of Arita ceramics prominently greet visitors once they arrive. Besides displaying ceramics made by local artists and residents at the train station, *hama*s (a tool used for porcelain packing process) were also given out as free souvenirs. This served two purposes. It connected tourists with the rich porcelain-making history of the town by giving them a free keepsake and also recycles industrial products as tourist souvenirs. The pavements in the paths leading right up to Izumiyama in Arita and walkways around the Imari train station are literally lined and paved with Arita porcelains. Throughout both towns (Arita and Imari), posters were also put up to indicate the latest exhibitions that were going on in town. They were informative to both foreign as well as Japanese tourists, informing them of the latest exhibitions.

Expressions of local identity could also be found with souvenirs in the form of Arita ceramic bowls holding local curry sauces sold at the train station and local stores for tourists interested to take back a local cuisine as well as the town's iconic ceramic artifact. Other unique souvenirs include handkerchiefs with classical Arita/Imari designs. Sometimes, Arita's local identity is reflected in national celebrations. For example, my fieldwork in Arita in the summer of 2013 coincided with Mount Fuji's success in attaining the World Heritage Site recognition. Arita's Fukagawa, which uses the Mount Fuji trademark as its company logo, featured a series of products to celebrate this momentous occasion. Coordination with national campaigns (e.g., applying for World Heritage status) provides an additional layer of social legitimacy to Fukagawa as a national brand in addition to its status as a purveyor of the imperial household in the sector of ceramics. Sometimes, Arita's parochial identity within Japan is projected outwards through diplomatic and foreign relations-related initiatives. Foreign dignitaries visiting Arita were also accorded special treatment, leaving behind their personally hand-signed works of art. Notable visitors were invited to sign their names or create special designs for display at the Ceramic Research Center.

Similarly, in Imari, the town's close association with porcelain art is reflected in the diverse offerings of souvenirs for visitors. Dried cuttlefish, which Imari and the prefecture of Saga is famous for, is wrapped in paper folders with drawings of lidded Imari-style ceramic vases and embossed with the words "VOC." Sets of *otsumami* (snacks that go along with alcohol, e.g., dried tempura and *aji* fish) are also wrapped in porcelain-themed packaging for sale as souvenirs for tourists. In the hotel which I stayed in, the restaurants were named after different porcelain genres (for example, *seiji* or green wares). My hotel room key card also has a picture of a Nabeshima porcelain design and the express checkout key box is lined with mini porcelain tiles with the *kotobuki* (longevity) designs.

CONSUMPTION

The legitimacy of Arita products comes mainly from the public's association of the town with its long history of ceramics ware. Historical legitimacy is backed up by artifacts displayed in the showrooms and galleries of well-known ceramics makers, museums, governmental promotion, exhibitions and the recorders of the town's history. Well-known brands like Koransha and Fukagawa are further legitimized in their reputations by the fact that they are suppliers and purveyors of fine bone china wares to the imperial household of Japan. Some of the most important material evidence of the close association of Koransha wares with the imperial household are the drawings/plans of the ceramic vases presented to the Japanese Emperors since the Meiji period (1868–1912).

The main consumers of high-end ceramics like Fukagawa and Koransha are local consumers in Japan. Their consumption sustained the continued production of Arita wares in the region. Consumption is sometimes bundled up with identity. For example, together with my informants, I dined at a local Arita restaurant which served famous local product (or *meibutsu*) chicken meat in compartmentalized white Arita ceramics and English tea served in teacups of one's choice. There are many to choose from, the most expensive teacup was said to be valued at nearly US$2,000, according to my informant. Such cafes served the dual purposes of promoting local cultural identity as well as innovative ways to market the town as a tourist site. Tourism is a main source of consumption in the town. Upon arriving at the Arita station, a tourist can rent a bicycle to ride around town for the whole day for only 500 yen. One-hour guided tours are also offered at the Arita train station: 1,000 yen for 1-2 individuals; 1,500 yen for 3 to 5 individuals and 3,000 yen for 6 to 15 individuals. At the train station, brochures have been translated into English, Chinese, German in addition to Japanese-language ones.

In the community, there was little dichotomy between tourism promotion, local economy integration and individual artistic expression. As part of the community, there was a role for everyone to play. The promotion of Arita of its tourism and its products were a single comprehensive package. The Arita Chamber of Commerce and Industry plays an important role in this. When I was in Arita, the institution was busy planning the activities for the 400th year anniversary of Arita. Together with cultural promotion, consumption by visitors, tourists and collectors drove and sustained production. Both processes were mutually dependent on each other. Other than individual consumers and tourists, Arita producers also depend on commissioned works for income.

ENERGY USE

The central position of energy use in firing ceramics is visible throughout Arita. Even at the train station, visitors are able to observe the contrast between an unfired and fired ceramic vessel. In the Kyushu Ceramic Museum, the evolution of energy sources is laid out clearly for visitors to view from timber-fired *noborigama* (climbing) kilns to coal and oil and finally, gas kiln. Public awareness of the importance of energy use is displayed at the tourist information booth at the Arita train station and a commercial restaurant near the station. Resembling the two-chamber beehive kiln in a photograph display at the Kyushu Ceramic Museum, I came across a similar two-chamber kiln in the Arita College of Ceramics that uses timber pieces for firing. There were contemporary efforts by dedicated pottery-makers to construct *noborigama* climbing kilns but some of them were closed down to make way for other uses of the landed property. Land use to fit the size of the climbing kiln is a major challenge for any land-scarce community. The issue of size is resolved by the use of gas kilns and especially electric kilns that are compact, stand-alone and sometimes portable on wheels. Some of the smaller gas and electric kilns can be operated by only one individual alone.

Coal was the main fuel in the age of modernization for Arita kilns, especially for well-known Arita brands like Koransha after the Meiji Restoration of 1868 as well as other kilns engaged in industrial production. While coal was more efficient for combustion than timber, it also produced large quantities of black smoke and soot, causing pollution. Pollution was an important issue which ceramics makers had to contend with. By the 1960s, with increasing environmental awareness in Japan, there was rethinking about fossil fuels used for firing and gas emerged as a cleaner alternative. Gas kilns represented the cleanest fossil-fuel based kiln technology available conventionally. There are little or no emissions from the firing process, with the exception of some minor discharge of water in some cases. Production-wise, for certain genres of ceramics, a veteran Koransha technician informant told me that gas kilns with their precise temperature controls were optimal for firing white wares due to their sensitivity to temperatures.

Energy use is closely related to the environment. One of my informants noted that there was a time (postwar 1960s) when she heard older residents mentioning about the need to avoid drying white linens and clothes in the open as they may get sooty from the emissions released from coal and oil use. According to the text of the explanation boards found at the Ceramics Research Institute, the shift from oil to gas kilns in Arita took place in 1965. This coincided with a period of environmental awareness in Japan. Gas and eventually electric kilns dominated Arita's pottery-making studios due to their environmental-friendliness and ease of use and controlling temperatures.

ENVIRONMENT

Historically, there was a connection between the producers and their immediate environment, with timber coming from nearby forests for firing, clay from Arita's own kaolinite quarry (Izumiyama) and coal from surrounding areas. But this has changed as Arita potters now get their supplies from sources external to Arita: clay from Amakusa for example, and firing is currently carried out using gas kilns without the need to feed wood fuels. In terms of the general environment, there are efforts to keep the public spaces clean and tidy in both Arita and Imari. Trash cans (or dust-boxes) at all drinks vending machines and other public spaces have separate sections for trash made of different materials. Burnable and non-burnable trashes are also separated. Inside the building, the flush toilets in both Arita (at the train station for example) and Imari had social campaign messages to remind users to conserve water resources.

Other than procedures to keep the physical environment clean, creative designs on ceramics products feature images drawn from nature, including animals and plants. These porcelain wares are used to craft manmade microcosms of natural spaces, e.g., ceramic basins, urns and bowls to rear goldfishes. Aside from the production matters, environmental themes are integrated into the conceptual idea of the products themselves. Nature is sometimes condensed and essentialized into a crafted space, miniaturized and made to resemble the natural environment, either in a fish bowl or a rock garden.

Public gardens that integrate ceramic arts with the natural beauty of the environment are carefully-crafted with two aims. In the pragmatic aspect, it is a marketing tool that draws domestic visitors and foreign tourists to consume porcelain art pieces and utilitarian wares in a comfortable environment. Visitors are able to compare and contrast the beauty of fired clay with the aesthetics of the Zen-like crafted gardens built around these consumption spaces.

In the aesthetic aspect, even if the pieces are mass-produced utilitarian wares, their designs when embedded inside a crafted natural environment, convey a sense of naturalness and negate the artificiality of manmade crafts. In my observation fieldwork, one of my informants (a skilled veteran technician at a top-ranking ceramics production facility) told me rather apologetically much of the craft-making process in making utilitarian wares which has been mechanized as part of a rationalized factory process, but he was grateful that some manual inputs continue to be important (e.g. hand-painted stylized bordering and drawings on higher-quality utilitarian ware pieces). The resident potters of Imari continue to operate a timber-fired *noborigama* (climbing kiln) out of respect for traditions and to keep the skills alive in circulation. Messages urging environmental awareness are written on slogans displayed at Arita's public facilities. Similarly, in Imari, environmentally-friendly practices can be found in local hotels and public spaces. Because of the efforts of the community in keeping the environment clean, I was able to spot storks, herons and cranes co-existing with Arita and Imari residents in the rice fields, residences and even near downtown's retail sector.

SURROUNDING REGIONS IN KYUSHU

It is the combination of human talents, creative environment and soft skills infrastructure that facilitated the two regions of Arita and Imari in maintaining their status as premiere pottery-making towns. Arita and Imari are also located in a prefectural macro-region where there is an active fascination with ceramics. In Saga prefecture, other cities and towns showcase the aesthetics of ceramic art forms. Saga city, the prefectural seat of the regional government for example, has a healthy enthusiasm for ceramics art works.

Besides Saga City, Karatsu is another neighboring city also engaged with ceramics-making. Karatsu is sometimes classified as a type of proto-porcelain with its glassy high-fired stoneware texture and so, it is sometimes regarded as the forerunner of Arita/Imari porcelain. The skills needed in firing proto-porcelain are applicable and useful for firing Arita/Imari porcelain at even higher temperatures in Japan's ceramics developmental history.

Other than Arita and Imari themselves, even the largest city in Kyushu integrates Arita works into its infrastructure. The main consumption, finance and business center in Kyushu is the city of Fukuoka. Here, there are also some very public displays of ceramics art in areas with a heavy pedestrian presence. The Hakata train station for example has large wall and column displayed of Arita tiles painted by ordinary residents of the city.

CONCLUSION

In writing about the ecology of the towns of Arita and Imari, I tried to characterize their typology as physical entities. They are not large urbanized cities like those described by Richard Florida in his work on the creative class.[11] Unlike his focus on the creative individuals, I was more interested in how the community functioned in producing economic outputs and, based on the creative output of the two towns, the major non-agricultural products made in Arita and Imari tend to be ceramics-based projects rather than university services, manufacturing advanced technologies or industrial products that Florida was concerned with in his study of the economies of international cities. I was also less concerned with issues about harnessing global talents which are the preserves of the world's leading metropolitan cities. Foreign and local talents in Yingge, Arita and Jingdezhen are specialists in ceramics making, rather than talents in all walks of life.

Some points from Florida's writing that may be applicable to Arita and Imari are discussed in the following paragraphs. The idea of agglomerations and clusters mentioned in Florida's publication[12] is applicable to the Arita/Imari cases. Gathering together the top porcelain-making firms (over a long historical period starting from the origins of the town itself) gather skilled talents within and outside the two towns to share ideas in ceramics-making and contribute their individual expertise to the collective. Third, the idea of social capital mentioned briefly by Florida in the engagement of community residents in their community institutions[13] is less applicable to a tightly-knit community or hamlet. However, the annual Arita sale is an opportunity for individuals to become stakeholders engaged to play different roles at different levels of the project. This prevents community residents from being completely alienated from each other. There are economic incentives involved, given that the annual sale is the most profitable sale for majority of the retailers. The central organizing institution in this case is the Chambers of Commerce and Industry, working for community (economic) sustainability and camaraderie during this period.

Given the large gap between my case studies and the needs of international metropolitan cities, I turned to studies of smaller communities. Amongst the literature that I reviewed was Brian Moeran's study of the Sarayama pottery-making community.[14] Moeran's Sarayama community resembled Arita and Imari more than the large cities depicted in Florida's publication, especially in terms of scale and function. But comparison with this community came with its own set of challenges too. I will discuss five points here. First, Sarayama is organized like a hamlet with close interconnecting ties between community residents. By comparison, my informants at the Folklore Museum of Arita mentioned that cooperation and networking tends to be based on projects. Second, because Sarayama is organized like a hamlet, its production is based on collective work in cooperatives. Arita's smaller studios may engage in some form of cooperative work but its large utilitarian ware producers are stand-alone institutions with world-class in-house facilities for research, marketing and distribution functions.

Third, because Sarayama potters produce *mingei* or folk wares, traditional methods of production are retained with an emphasis on individualistic artistic creation. In Moeran's account, water-powered clay crushers, timber-fired kilns and traditional equipment used in pottery-making such as the manually-operated potter's wheel continue to be in use. These technologies have become museum artifacts or relics in Arita and Imari which have invested in more advanced technologies like gas kilns. In Arita and Imari, the use of traditional kilns has completely disappeared except for symbolic purposes. For Arita, mechanization and progressive use of cleaner fuels were important for producing high-quality fine bone china porcelain that required precision in firing with greater productivity and lower levels of defects. Larger volumes of porcelains can be fired at higher temperatures compared with *mingei* products which tend to be fired at lower temperature and have rustic (*shibui*) appearances and in smaller quantities. In other words, Arita/Imari and Sarayama are producing different types of products. Moreover, Arita has world-class technologies and research facilities that work closely with the porcelain makers like the Ceramic Research Institute and the Arita College of Ceramics. These research institutions offer world-class technological facilities and knowhow for shared projects with smaller studios that do not have the same finances resources as their larger counterparts.

Fourth, the Sarayama community serves mainly a niche consumer base, described in Moeran's Chapter Seven on the *mingei* boom.[15] The bigger producers in Arita, on the other hand, have a bigger international customer base that is less dependent on domestic trends. Moreover, due to the historical legitimacy of Arita and Imari products, there are established collectors' communities both in Japan and overseas keen on collecting such products. Fifth, the commonality between Arita/Imari and Sarayama products is found in their hybrid East-west fusion of influence. In Moeran's case, he characterizes it as a fusion of "European medievalism and orientalism."[16] Globalization and rapid exchanges between cultures and civilizations has rendered any attempts at identifying original inspirations for art works and designs meaningless. In Arita's case, my informants preferred to characterize their products as "Japanese-style" rather than oriental, Asian or Western. After many cycles of influence, first from continental East Asia and then the Europeans, Arita/Imari designs are unique after indigenizing and drawing inspirations from other cultures. In fact, Japanese Arita/Imari ceramics designs have become influential on European potters and also went on to influence ceramics-making in continental East Asia.

NOTES

1. I acknowledge the help rendered by the potters, ceramicists, stakeholders, firms in Arita and Imari as well as the prefectural government of Saga. Because of introduction and the personal networks of stakeholders, I was able to visit personal collections and other non-public access locations. I gratefully acknowledge their generosity. I am keeping the identities of all my informants confidential to protect their privacy.

2. In some high-end commercial products, porcelain decorated and designed in the classical Imari-style export wares of the seventeenth century are known as *ko-Imari yoshiki* (Old Imari style).

3. For readers interested in looking at photos of European aristocratic spaces decorated with Imari porcelains, please refer to: Ōhashi Kōji, *Umi o watatta koimari — seramikku rōdo* [Exported old-Imari ceramic road] (Kyoto: Seigensha, 2011). Ōhashi is a consultant for the Kyushu Ceramic Museum.

4. Fukagawa Tadashi, *Koimari no bi to roman* [Old-Imari's beauty and romance] (Tokyo: Shufunotomo, 1979), 22.

5. Kyodo, "Arita Kicks Off Annual Ceramics Fair," *Japan Times*, April 30, 2001, accessed July 31, 2013, http://www.japantimes.co.jp/news/2001/04/30/national/arita-kicks-off-annual-ceramics-fair/#.UfkWvV_yyUk.

6. Preparation of Amakusa clay for firing is outside the scope of this chapter. For more ideas in this area, readers can refer to: Association for the Promotion of Traditional Craft Industries, "Amakusa Pottery and Porcelain," accessed August 1, 2013, http://kougeihin.jp/en/crafts/introduction/pottery-and-porcelain/2867?m=pd.

7. According to the plaque at the site of Izumiyama Jiseki-Ba (Izumiyama Kaolin Quarry), kaolin or kaolinite is a liparite (rhyolite volcanic rock) containing mainly quartz and sericite with some whitish-colored iron.

8. See for example, Yoshioka Yasunobu, *Tōjiki no sekai* [The world of porcelain] (Tokyo: Yamakawa Shuppansha, 2001), 63.

9. A detailed discussion of Japan's ceramics trade with East Asia is found in an earlier chapter of this book. Readers who are interested in a specific port case study such as Hakata in such trading activities can also refer to the informative publication: Ōba Kōji, *Chūsei Nihon saidai no bōeki toshi* [The biggest trading city in Japan during the Medieval Period] (Tokyo: Shinsensha, 2009). I gathered this literature resource from my fieldwork at the Fukuoka City Museum. Readers may also be interested in a new English language publication that was published at the end point of this writing which talks about the influence of China on early modern Japan's porcelain production. It was recommended by an anonymous reviewer: Nicole Coolidge Rousmaniere, *Vessels of Influence: Chinese Ceramics Imported into Japan and the Formation of the Porcelain Industry* (Bristol: Bristol Classical Press, 2012).

10. In one of its unmarked showroom display captions, the Ceramic Research Institute defined "Imari" as a location in the Hizen region during the Edo period (with Arita in its center) where porcelains were made. Its port exported manufactured products like ceramics, dyed textiles and vases to the rest of Japan.

11. Please refer to: Richard Florida, *Cities and the Creative Class* (New York: Routledge, 2005).

12. Ibid., 29.

13. Ibid., 30.
14. Please refer to: Moeran, *Folk Art Potters of Japan*
15. Ibid., 140–55.
16. Ibid., 221.

Chapter Ten

A History of Jingdezhen Kilns, Its Development, and Energy Use (with Comparative References to Arita)

The success of Jingdezhen's porcelain industry is directly related to the availability of natural resources the town has around it. Wood was available in proximate areas[1] around the town for fueling kilns to fire porcelain wares. Wood combined with a large workforce and large domestic and export markets became comparative advantages for Jingdezhen's pre-modern historical development. These advantages were only mitigated by political and dynastic changes within China. For at least over five hundred years until the late 1950s, Jingdezhen relied on timber for firing its products even as neighboring Meiji Japan gradually shifted to coal energy in the early twentieth century following the advice of European technicians. Coal-firing was introduced in a substantial way approximately 150 years later in the People's Republic of China's ideological and policy move towards rapid heavy industrialization. Besides fuel, Jingdezhen's advantage in possessing kaolinite resources remained before it was gradually eroded by the advent of modern European chemistry which succeeded in producing kaolinite alternatives by the early eighteenth century.

This chapter discusses two important continuities in Jingdezhen's history. Throughout the alternating historical periods of political turmoil and periods of peaceful stability as well as competition from the Europeans and Japanese, Jingdezhen remained as an important ceramics-production center which turned out two major lines of products—mass-produced affordable utilitarian and export wares as well as high-quality artistic and "imperial" wares (and their contemporary reproductions). The adaptation of new kiln technologies and corresponding use of new fuels sustained the Jingdezhen kilns' relevance to the ceramic production industry for more than half a millennia. The social history of the kiln technology usage and the energy transitions experienced by the kilns are discussed in this chapter.

LITERATURE REVIEWED IN LIBRARY RESEARCH

Before going on to fieldwork in Japan, Taiwan and China, I carried out library research at the Chinese University of Hong Kong (CUHK). Using secondary sources, I reviewed scholarly writings on the historical origins of the Jingdezhen kilns. Amongst the publications found in Hong Kong, Geoffrey C. Gunn's publication probably has one of the most succinct descriptions of the historical background of the Jingdezhen kilns. According to Gunn:

> …Jingdezhen in present-day Jiangxi province was for centuries the major global ceramic production and export center. Dating from the 6th century CE, its status as a production center was consolidated under the Song. By Ming (1368–1644) times, production of high-quality porcelain at Jingdezhen surged; it became the procurement center for the royal household, and also met the needs of the thriving export market. Owing to local disorder, production stopped in 1506–21 and again in 1567–72, when the government orders failed. Jingdezhen was also temporarily closed in 1650–80 during the Ming-Qing transition. Under the Qing (1644-1911), Jingdezhen furnaces were revived as "imperial kilns," under the close attention of the emperor and high officials.[2]

Three points are important in this narrative. First, Jingdezhen production is dependent on political stability in China. Therefore, innovation, mass production and high quality output were only possible during long periods of sustained political peace and order. Second, at last from the time of Song dynasty, Jingdezhen was already engaged in voluminous production for the domestic market in China. Therefore, Jingdezhen's experience in mass manufacturing started early from at least Song dynasty. Because of its large-scale operations and massive use of raw materials, Jingdezhen needed to grapple with resource supply and manpower needs from this period onwards. Third, quality has always been important, driven by customer needs, both in terms of overseas and domestic consumers. I will discuss three historical periods of political instability affecting Jingdezhen ceramics production below.

MING-QING TRANSITION

At the peak of Jingdezhen porcelain production during the Ming dynasty, political instability due to incoming Manchurian incursions resulted in one of the most significant disturbances to Jingdezhen's production schedule. However, the reconstruction of the kilns after that resulted in yet another prosperous era for the Jingdezhen kilns in the Qing dynasty. In the publication *Porcelain for Palaces: The Fashion for Japan in Europe, 1650–1750* written by John Ayers, Oliver Impey, J.V.G. Mallet and published by the Oriental Ceramic Society in London, I spotted several photographs of ceramics made just after the Jingdezhen kilns were painstakingly reconstructed in the 1680s to restart mass production when the fighting between the Manchu conquerors and Ming resistance crumbled.[3] The volume published rare pictures of Jingdezhen wares imitating Japanese Kakiemon and Arita-styles ceramics as Jingdezhen tried to recapture European markets previously lost to the Japanese kilns during the volatile Ming-Qing interregnum period.[4]

LATE QING DEVELOPMENTS

My textual research work in CUHK Art Museum yielded other materials on specific periods of history in Jingdezhen when political events affected the kiln production. During the late Qing period in the reign of Emperor Guangxu, the impact of the Taiping Rebellion and its impact on kiln operation were narrated in one of the Museum's older publications. The Qing Imperial Kiln at Jingdezhen was destroyed during the Taiping Rebellion in 1855 and it was not until the Guangxu reign that the Imperial factory had fully recovered and reached another peak in production in terms of both quantity and quality.[5]

This again indicated Jingdezhen's delicate dependence on external political stability for optimal ceramics output. China's struggles with its last dynasty contrasted sharply with Japan's relatively bloodless Meiji Restoration of 1868 when the latter modernized its manufacturing industries along European lines. Less than 50 years later, Japan's pottery-making Arita and Seto regions became the sites of modern ceramics production and gradually overtook the Chinese in implementing modern kiln technologies. At this point of time, the Japanese kiln operators were students learning from their teachers (the Europeans) who had gone through the Industrial Revolution. By the Guangxu era, European kiln technology surpassed the Chinese, their ability to produce high-quality standardized utilitarian wares was also superior and the European technicians had good knowledge of chemistry needed for the glazes and kaolinite alternatives. Because of scientific advancements, the Europeans were already producing large quantities of standardized western-style utilitarian dinner wares with relatively uniform quality and specifications which was the new benchmark in kiln technological sophistication.

The repeated political turmoil in China during the periods of Ming internal disturbance (1506–21 and 1567–72), the Ming-Qing interregnum and the Taiping rebellion gradually wore down the competitiveness and technological capabilities of Jingdezhen as European and Japanese potters and kilns caught up with Jingdezhen. Moreover, the Chinese needed to expend resources every time political changes destroyed the Jingdezhen kilns. But the need to constantly rebuild the kilns every time they were destroyed in the conflicts also meant that Jingdezhen enjoyed some form of creative destruction. Each new dynasty or political regime developed their own design preferences and the influx of foreign inputs from the Mongol Yuan to the Manchurian Qing and then foreign artists residing in contemporary Jingdezhen contributed to new forms, techniques and glazes in Jingdezhen-made products.

LITERATURE COLLECTED DURING FIELDWORK OUTSIDE HONG KONG

Besides the efflorescence of Jingdezhen wares in terms of styles and designs, the distribution of Jingdezhen products was also global in nature even from the pre-modern times. These exchanges along the overland and maritime Silk Road also facilitated exchanges in creative ideas about ceramics design and production, resulting in Western-style export ceramics made by Jingdezhen kilns. In my observation fieldwork in the Idemitsu Museum of Arts in Moji Japan, I came across the publication *Treasures along the Silk Road: Steppe and Oasis Routes/Sea Route*, which detailed the exchange of goods including porcelains through the land and sea routes in the pre-modern era.[6] I purchased a guidebook from Idemitsu Museum which showed a number of Jingdezhen wares present in this inter-regional trade.[7] There are altogether at least 51 items from Jingdezhen featured in the catalogue from the Yuan, Ming and Qing periods and most of them are blue and white porcelain. At least two of them were western-style teacups, probably exported to Europe while others were done in Western-style kitchenware shapes and forms embossed with Chinese designs or non-European makes with Arabian text.[8] I discussed this aspect extensively in my earlier chapter titled *Re-centering Trade Periphery through Fired Clay*.

OBSERVATION STUDIES – PRE-MODERN JINGDEZHEN EXHIBITIONS

Besides textual analysis, I also carried out observation studies. The display of fine Jingdezhen antique porcelains could be found in my fieldwork in Hong Kong's museums. Through these exhibitions and shows, I acquired knowledge about Japanese and Chinese export ceramics in major exhibitions or sales in the territory. Perhaps one of the most important exhibitions that I missed in Macau but caught in Hong Kong was the Maritime Porcelain Road exhibition. Here I was exposed to a variety of export porcelains, including the highly representative blue and white *kraak* dish with stylized designs dating back to the Tianqi period (1621-1627) of the Ming dynasty. This plate was typical of the Zhangzhou plates that were exported in large numbers and were sometimes known as "Swatow wares." Another major exhibition where I carried out observation studies in Hong Kong was the *Cruising the Universe: Fantastic Animals in the Arts of China* exhibition that displayed a number of Jingdezhen items such as a fish bowl with carps and goldfish designs done in *wucai* (five colors) design attributed to the Ming Wanli period (1573–1620).[9] Here I also observed another jarlet dated to the 1600s carrying the design of a young person riding a *kirin* also in *wucai* colors. Another Jingdezhen ware in the same show was a vase with dragon-shaped ears/handles and stylized enamel designs juxtaposed against a yellow background dated to the Qianlong rule of the Qing period (1736–1795). Outside the special exhibitions, the permanent display of the Hong Kong Museum of Art features some Jingdezhen pieces like a Yongle (1403-1424) blue and white large-sized dish with lotus and floral designs. In the narrative of the official brochure for the display titled *Gems of Chinese Ceramics* from the Hong Kong Museum of Art, however, accent was not surprisingly placed on the far southern region of China, including Guangdong kilns like Shiwan. The brochure did mention about Ms Lisa Ng's donation of Jingdezhen figurines in 2003 to augment the Museum's collection of mid to late twentieth century ceramics artifacts.[10]

I also carried out observation studies at CUHK Art Museum exhibition on Ming Jingdezhen porcelains in August to December 2012. Some prominent Hong Kong-based collections were featured in this exhibition including the Tianminlou and Huaihaitang Collections. This was quite a unique show, given that it was the first time that the Jingdezhen Imperial Ware Museum was willing to jointly run an exhibition with another institution (combined with The CUHK Art Museum (2012)'s own mid-fifteenth century Ming pieces). According to the narrative in the exhibition materials, Jingdezhen ceramics reached "unprecedented achievements" in the Ming period. Other scholars prefer to look at the Qing period as the epitome of Chinese porcelain development. According to Stephen Little, during the Kangxi period (1662–1722) of Qing dynasty, Jingdezhen perfected the brilliant red monochrome (*langyao*), oxblood red colour, peachbloom glaze and the *famille verte* overglaze enamel.[11]

Despite differences in opinion, most works almost universally agree that the Ming and Qing periods represented the most important developmental phases of Jingdezhen in terms of porcelain production. Because the CUHK exhibition displayed an imperial ware collection, I was able to observe physical artifacts representing the finest precision firing works done in the premodern period, i.e., kiln-firing using timber fuels by top craftsmen. The Huaihaitang Collection later became the subject of an exhibition by itself in the Art Museum of CUHK in end 2012. In the narrative found in the official brochure for the exhibition, the Museum writer indicated that the founding of the Imperial Porcelain Factory in the early Ming period led to "the maturity and standardization of production."[12] In other words, energy fuel use and the kiln-firing technologies in Jingdezhen and probably other parts of China used the imperial kilns as the benchmark of quality in firing porcelains. Table 10.1 is a list of exhibitions and shows in Hong Kong where I carried out observation studies.

My attendance at these exhibitions and reading about them in the media in Hong Kong gave me important insights into three areas. First, I was able to view pre-modern Jingdezhen products on display. This was important because I was able to tell the difference in texture, quality and designs between imperial and folk wares. I was able to distinguish high quality wares made in the pre-modern period versus those made in the contemporary era. Likewise, I was able to compare mass-produced pieces for export with utilitarian wares made in the contemporary period. Second, I could appreciate how large number of imperial wares had to be produced in order to sieve out the defective ones (or lower quality items) for destruction. By comparison, precision temperature controls in contemporary kilns are able to lower defect rates for manufactured pieces. The material artifacts themselves were therefore visual evidence of the chronological development of kiln-firing technologies in the pre-modern era. The products made by the imperial kilns served as benchmarks of quality and design and export wares were an economic indicator of global demand and supply. Third, these exhibitions demonstrated that Jingdezhen production was not only driven by design innovation and economics but also prevailing political conditions within and outside China. Political transitions negatively affected production output but periods of peace and stability saw kilns reconstructed quickly to meet renewed demand again.

Uniformly across these exhibitions, imperial wares were seen as the universal benchmark of quality. This is something different from the Japanese Arita case study where the Nabeshima kilns in Imari were probably the closest porcelain equivalent to the Jingdezhen imperial ware system, although they are different in many ways.[13] The Jingdezhen imperial ware system emerged in the Ming period but really only became highly-developed during Qing dynasty and it served as a benchmark for porcelain-making throughout the Chinese empire under the Manchus. The Nabeshima kilns in Imari, on the other hand, were older than the Qing imperial ware system and served a limited *han* domain audience. In my fieldwork at Arita's Kyushu ceramics museum, my informant mentioned about folkware or common utilitarian pieces that were eventually spotted and adopted as Nabeshima designs. These wares ceased to exist as ordinary Imari wares and attained Nabeshima status. I could not locate an equivalent system in Jingdezhen in my fieldwork

Table 10.1. Exhibitions and Shows that the Author Attended for His Observation Studies

Name of exhibition or event	Type of porcelain on display	Source
Rishi Taoci Canjuzhan	Japanese-style porcelain and tableware	Citistore, "Rishi Taoci Canjuzhan" [Japanese-style porcelain and tableware exhibition], *Headline Daily*, October 24, 2013, 35.
Fantasy World — Chinoiserie	Export ceramics from China	Fieldwork
Maritime Porcelain Road – Relics from Guangdong, Hong Kong and Macao Museums	Export Ceramics	Hong Kong Museum of Art, *Maritime Porcelain Road – Relics from Guangdong, Hong Kong and Macao Museums* (Hong Kong: Leisure and Cultural Services Department, 2013).
Cruising the Universe: Fantastic Animals in the Arts of China	Ceramics with animal and mythical creatures designs	Hong Kong Museum of Art, *Cruising the Universe: Fantastic Animals in the Arts of China* (Hong Kong: Leisure and Cultural Services Department, 2013).
Permanent Display: Gems of Chinese Ceramics from the Hong Kong Museum of Art	Variety of Chinese ceramics types	Chinese Antiquities Unit and the Design Unit, Hong Kong Museum of Art, *Gems of Chinese Ceramics from the Hong Kong Museum of Art* (Hong Kong: Leisure and Cultural Services Department, 2012).
Filling the Interregnum Ming Mid 15th Century Ceramics from Jingdezhen	Jingdezhen porcelains	Art Museum, The Chinese University of Hong Kong (CUHK), *Filling the Interregnum Ming Mid 15th Century Ceramics from Jingdezhen* (Hong Kong: CUHK, 2012).
Enlightening Elegance Imperial Porcelain of the Mid to Late Ming The Huaihaitang Collection	Mainly blue and white porcelains	Art Museum, The Chinese University of Hong Kong (CUHK), *Enlightening Elegance : Imperial Porcelain of the Mid to Late Ming , The Huaihaitang Collection* (Hong Kong: CUHK, 2012).

Chinese Ceramics and Seals Donated by the K.S. Lo Foundation	Chinese Ceramics and seals. Featuring a *meiping* vase in blue and white floral decoration dating back to Yuan dynasty (1271-1368).	Hong Kong Museum of Art, "Chinese Ceramics and Seals Donated by the K.S. Lo Foundation," *Hong Kong Museum of Art Newsletter* 2 (April-June 2013): unpaginated.
Flagstaff House Museum of Tea Ware	Permanent collection displaying a Qing blue and white imperial ware with dragon design on the 1.F Gallery	Flagstaff House Museum of Tea Ware, *Museum Guide* (Hong Kong: Flagstaff House Museum of Tea Ware, n.d.) (museum brochure obtained in 2012).
Fantasy World – Chinoiserie	Featuring French art with Chinese-inspired designs. Including blue and white pieces.	Le French May Organisation, *Le French May 17 April – 23 June 2013 French Arts Festival Hong Kong & Macau* (Hong Kong: Quality Printing Limited, 2013), 166-67.
Divine Power The Dragon in Chinese Art (10-11 Feb 2012)	Featuring artifacts with the Chinese dragon design including at least two Jingdezhen wares. A plate decorated with a green colored dragon dated back to the Ming Zhengde period and the other is a large eared flask with underglaze blue and red dragon designs dated back to the Qing Qianlong period.	Oriental Ceramic Society of Hong Kong Limited and Art Museum, The Chinese University of Hong Kong, *Divine Power— The Dragon in Chinese Art* (Hong Kong: The Oriental Ceramic Society of Hong Kong Limited and Art Museum, The Chinese University of Hong Kong, 2012).

or textual analysis. In terms of political and management control, the Jingdezhen imperial ware system was far more centralized and acted as powerful cultural symbols of artistic and creative inspiration for other kilns in China, compared with the Nabeshima kilns.

IMPERIAL KILNS

My observation studies in the Hong Kong Heritage Museum yielded additional information on the Jingdezhen imperial kilns during the Ming and Qing dynasties. According to the narrative displayed at the museum, official kilns which specialized in supplying the court with porcelains were established during the Ming dynasty. Products made in these kilns carried official marks and certain designs and glazes that became hallmark icons of imperial wares during this period. These glazes included monochromes, underglaze blue, Xuande red glaze, Chenghua doucai, Hongzhi bright yellow, Wanli *wucai*, according to the textual displays at the museum. Also, according to the museum's textual display, the imperial kilns reached their peak condition during the Qing dynasty, partly due to highly specialized division of labor and skills. The narrative in the museum argued that innovations in forms and shapes were possible due to denser and firmer clay bodies used for porcelain production. Michael Dillon's 1978 article contained statistics of porcelain wares fired in the imperial kilns between 1529 to 1571 and the expanding orders for uniformly high quality pieces created pressures for proto-industrialization and mass production in Jingdezhen during this period.[14]

Table 10.2. Statistics of Number of Porcelain Wares Fired in the Imperial Kilns between 1529 and 1571. Source: Michael Dillon, "Jingdezhen as a Ming Industrial Center," Ming Studies 6 (1978): 38.

Date	Number of Pieces Fired
1529	2,570
1531	12,300
1544	50,000
1555	79,750
1571	105,770

NON-IMPERIAL KILNS

Imperial kilns probably represented the zenith of kiln technology during the Ming and Qing period but many other kilns in Jingdezhen were operated by owners making utilitarian wares for daily use amongst ordinary folks. The first kilns in Jingdezhen were of the beehive type at the location's early industrialization phase.[15] Like the *katamaki*s (kiln hands) in Japan, Jingdezhen's kiln-firing workforce consisted of partially-skilled manpower hired on limited-period contracts with the job description of setting up the firing process, managing the fire and then put away the setup after use.[16] The scale of manpower involvement is not the only superlative feature of the proto-industries. Challenges encountered in handling fuels for firing large kilns were also detailed in the Stephen Little's article quoting Pere d'Entrecolles:

> Some one hundred and eighty loads of pine fuel (of a hundred and thirty-three pounds weight each) are consumed at every firing, and it is surprising that no ashes even are left...it is rare for a furnace to succeed completely, often everything is lost, and on opening it the porcelain and the cases [saggers] will be found converted into a solid mass as hard as rock...[17]

The naturally-occurring kaolinite in Jingdezhen China gave rise to the Jingdezhen craftmen's ability to control the firing process because, when kaolinite is mixed in different proportions with clay, the fired clay will react in different ways.

> This was a particularly useful principle for wares that were fired in the unusual Jingdezhen "egg-shaped" wood-burning kilns, used at this kiln site from the late Ming dynasty until the 1950s. These kilns (shaped like half an egg lying on its side and about fifteen meters long inside) showed a substantial fall-off of heat from their fireboxes towards their chimneys—from about 1320–1000 degrees Celsius within a single kiln. This unusual distribution of heat allowed Jingdezhen potters to fire thousands of porcelain pieces simultaneously to a wide range of temperatures, with a variety of glazes, all within a single firing. Plain white glazes for later enameling would be placed in the hottest part, followed by underglaze-blue wares, and then by high-temperature monochromes, such as celadons and cooper-reds....More than eighty such kilns were still in operation at Jingdezhen in the mid-1950s and a couple of wood-fired *zhenyao* kilns still operate in the city today. [18]

The description above indicated the sophistication of pre-modern Jingdezhen kilns by the time of the Ming dynasty. A single kiln chamber is able to accommodate differential levels of heat in different sections of the chamber to fire various glazes. This is energy-efficient because it meant that Jingdezhen kiln operators do not need to repeat firings for each and every genre of glazes with their differential temperature needs. One firing accommodates the entire range of products with their different glazes. This is probably the reason why the kilns remained in use until the 1950s. In addition, as the passage indicates, while the Jingdezhen imperial kilns are no longer operational, some older folk wood-fired heritage kilns (reconstructed or otherwise) in Jingdezhen continues to operate for symbolic heritage purposes or limited production of porcelain pieces. This is similar to the Fukagawa-owned *noborigama* kiln that I visited in Arita or its counterpart in Imari. These Jingdezhen kilns also served another purpose as tourist attractions and are promoted by attractive marketing materials as part of Jiangxi province's advertising campaign.

At the Hong Kong Heritage Museum located in Shatin, I was able to observe the Jingdezhen pre-modern production of porcelain from a series of watercolor paintings dated back to the eighteenth century and drawn from the collection of the university Library of Lund at Sweden. Here, the pictures chronologically depicted the processes of crushing kaolinite with hydropower-driven pounders, contraptions very similar to the ones that I saw in the town of Arita in front of the town's folk museum. Besides the use of hydropower, brute animal strength was also used to mix and crush the clay mixture. After that, the clay was washed. Woodcutters chop up the firewood for feeding the kiln. This division of labor can still be found in Japanese pottery-making hamlets like Sarayama but, in Arita, wood and clay now comes mainly from out-of-town sources, e.g., clay is imported from Amakusa. The chopped-up wood is then slowly fed into the kiln. After the firing is done, the finished products are then packed into saggars, ready for transportation to clients. Besides production issues, during my observation fieldwork in the Hong Kong Heritage Museum, I took notes on how underglaze blue became a defining feature of Ming Jingdezhen wares:

> By the time of the Ming dynasty, the technique of firing underglaze blue porcelain had fully matured in official and civilian kilns. Underglaze blue porcelain was the major form of porcelain production. The kilns in Jingdezhen in Jiangxi province not only provided the court with official porcelains (Text compiled from explanation display board at the Hong Kong Heritage Museum).

Some samples of Jingdezhen wares produced in folk kilns in the late pre-modern period (late Qing or Republican-periods) were encountered in my observation fieldwork in Japan. This historical period probably represented the last phase of porcelain production before the advent of modern kiln technologies and management techniques.

While Chinese imperial and industrial kilns had access to raw materials and manpower, the Europeans who were on the cusp of the industrial revolution began to develop efficient methods of firing porcelain and synthetically producing the bone china effect. Later European competitors not only caught up with Jingdezhen technological but produced cheaper wares more efficiently. Titus M. Eliens, Head of Collections at Gemeentemuseum Den Haag, made the following important estimation:

> It took about a week to fill, fire and empty each kiln. A kiln loaded with flatware could hold around 4,275 items. Therefore, assuming an annual production capacity of 52 kiln-weeks, each kiln could produce approximately 220,000 plates a year. A 1668 fire brigade inspection report says that there were 26 potteries in operation at that date, with a total of over 41 kilns between them. Assuming, therefore, that all the potteries produced only flatware, they could have produced a total of around 9 million items in that year.[19]

The passage above contained at least three important points. First, the Europeans were becoming more self-dependent on high-quality bone china wares starting from the seventeenth century. The European kilns used high-quality Jingdezhen and Arita wares as their benchmarks in designing the table wares, indigenizing styles and adapting firing techniques in the process. Over two hundred years later, with constant innovation and technological upgrading, the Europeans (such as German chemist and engineer Gottfried Wagner would be one of the leading names) would return to East Asia starting with Japan to teach Asians how to fire porcelain using modern coal-fired kiln. Second, these kilns were still pre-modern in nature, requiring one week of manual labor to prepare the kiln for firing. Third, the number appears large although there are no comparable figures on the Jingdezhen for comparative analyses. But it did indicate that Jingdezhen's competitors were able to produce ceramics in sizable quantities at least by 1668. The same article also mentioned that the Dutch Blue delft porcelains with the white tin glaze imitating kaolinite was made only at about one tenth of the prices of equivalent Chinese porcelains.[20] Gillette's important work on contemporary Jingdezhen quotes Kerr and Wood in tracing back Jingdezhen products' decline in the European market to the time when the German, French and English kilns produced authentic porcelain starting from 1709 when Meissen potters mixed kaolinite with calcined alabaster flux.[21] Previous to this development, many different kinds of imitation materials were used, but they were not considered as genuine porcelain like those made purely with kaolinite.[22] The Meissen version came closest to real porcelain texture and feel. My fieldwork in Arita's Kyushu ceramic museum also indicated that Japanese potters received technological transfers in the late nineteenth and early twentieth centuries when the Europeans taught them how to make synthetic blue glaze in the lab using cobalt for blue and white wares. The Industrial Revolution diminished the importance of geographical proximity of natural resources to Jingdezhen because these materials could not be created in the labs.

Besides using mass production techniques and scientific methods to create kaolinite-like substances, the Europeans also adapted and indigenized Jingdezhen and Arita designs effectively. I also collected museum catalogues that contained ideas about the influence of Jingdezhen wares on Japanese porcelain. The Saga prefecture's Kyushu pottery cultural museum collaborated with The Hague City's art museum and Amsterdam national museum of art to produce a volume on *Arita Delft China*. The advantage of combining the resources of all three museums facilitates comparisons between Jingdezhen wares and Arita as well as Dutch Delft wares. From these comparative studies, it was possible to understand how the Dutch were able to closely imitate and then indigenize the designs even as they utilized better firing technologies to lower the costs. Page 23 of this publication—*Arita Delft China,* has a comparison between a 1590-1630 Jingdezhen plate and a 1700 Delft ware with similar designs (water-based birds drawings).[23] Overall, the net effect of scientific development and technological advancement was that the Europeans were beginning to gradually produce more products that resembled Jingdezhen and Arita designs but with indigenous flavor and using synthetically-produced chemicals and more efficient methods of mass production. Eventually, Western kilns transferred their knowledge to Japan and initiated the kiln modernization in East Asia. Jingdezhen's own conversion to coal came after the Second World War in the early heavy industrialization phase before private enterprises introduced the latest models of gas-fired kilns to the porcelain-making town. In the next section, I will discuss Jingdezhen's postwar contemporary development.

CONTEMPORARY JINGDEZHEN

One of the most informative and updated articles I located on the subject of energy use in contemporary Jingdezhen was Maris Gillette's article "Copying, Counterfeiting, and Capitalism in Contemporary China: Jingdezhen's Porcelain Industry" published in *Modern China*. Maris Gillette's updated work on Jingdezhen provides a brief history of energy use in Jingdezhen. I have tabulated this history in table 10.3.

This energy transition is consistent with the general evolutionary path of energy use in other industries that burn fuels. Most economies and industries

Table 10.3. Brief History of Energy Use in Jingdezhen. Source: Gillette, "Copying, Counterfeiting, and Capitalism in Contemporary China," 33. Gillette also quoted from this source in discussing about the timber supply industry: Lili Fang, Jingdezhen minyao [Jingdezhen's folk potteries] (Beijing: Renmin meishu chubanshe, 2002), 179-81.

Year	Energy use in Jingdezhen
Before 1958	Jingdezhen uses wood and bramble kilns. The traditional phase of porcelain-making in Jingdezhen.
From 1958 until 2002	Central government decreed Jingdezhen porcelain fired in coal-powered kilns. Result: wood cutters and firewood distributors went out of business and closed down. Marks the phase of heavy industrialization and mass production of porcelain wares in contemporary Jingdezhen.
2002	Private sector businesses imported gas kilns into Jingdezhen. They come in various sizes, capacities, technological sophistication and functions. The Chinese government then banned used of coal to manage air pollution, carbon emissions. Marks the beginning of the environmentally-conscious phase of porcelain production in Jingdezhen.
September 2004	Coal-fired kilns closing and all are completely closed down by 2005. The start of the natural gas phase.

follow this evolutionary path because of the following reasons. I will briefly mention about general characteristics of energy use for contemporary manufacturing industries before discussing more specific features of the ceramics industry further below in the text. First, the evolution from wood to gas becomes increasingly environmentally-friendlier. Given most developed (and fast-developing) societies' stated policy directions for a better environment, the use of a clean fuel with little carbon footprints like gas is a logical option. Natural gas is also easier to transport than solid fuels like coal and liquid-based fuels like oil. Second, scientific rationalization promotes greater productivity and less wastage (lean production) and this is true for almost all manufacturing industries in general. In the case of the ceramics industry, fuels that promote precision controls for controlling temperature are ideal in minimizing defective products. I will discuss more about this below. Third, rationalization of industries tends to promote better efficiency, evidenced by the disappearance of the timber supply industry. Cutting down on supporting industries and intermediaries in transportation of raw materials also reduce costs and wastage—both of which are hallmarks of industrial rationalization.

As the next in line of fossil fuel use, Gillette describes the gas transition in energy use in Jingdezhen succinctly:

> Most porcelain in Jingdezhen was fired in gas (propane) kilns, but some of the finest antique replicas were fired in wood-burning kilns, to imitate more closely the process used in imperial times. In 2005, there were three wood-burning kilns in Jingdezhen, all licensed by the government. There were also wood-burning kilns in southern Anhui province, and some Jingdezhen producers transported unfired ceramics there.[24]

This passage is revealing about the energy use situation in Jingdezhen. First, only three timber-fired kilns remain in the whole of Jingdezhen and all regulated by the state.[25] This number is approximately similar to the number of wood-fired kilns that I encountered in Arita and Imari Japan. The main difference, however, is that the wood-fired kilns in Arita and Imari are privately-owned in the case of Fukagawa and operated as a cooperative collectively by the residents of Imari's Okawachiyama pottery-making district. In the case of Hong Kong, according to my informants, the government has completely ceased all use of wood-fired kiln due to reasons of pollution amongst others. Japan has also gone through the phase of pollution caused by coal use during the fast-growth period of the 1960s. For example, the Japanese pottery-making regions of Seto and Arita once experienced soot settling down on the town's residences due to carbon emissions from coal use. Jingdezhen's switch to gas kilns after 2005 also reflects environmental concerns. On the other hand, there are still some limited artistic applications for the traditional timber-fired kilns. Despite priorities in industrial rationalization, environmental concerns and quality issues in output, there remain specific uses for wood-fired kilns. For example, Maris Gillette's article pointed out the usage of wood-fired kilns in Jingdezhen for a specific creative function – to make reproductions of imperial wares from the Ming or Qing period for sale to collectors and clients.[26] Because imperial wares were made in China's pre-modern history (specifically Ming and Qing periods), they were fired in kilns that used timber and traditional methods of production. These procedures need to be replicated for making authentic imperial wares. These fine reproduction wares are sometime dispensed as gifts to foreign dignitaries or popularized in Chinese cultural products.

Second, there are similarities in both Jingdezhen and Imari/Arita that resident craftsmen try to imitate past Jingdezhen designs and patterns. According to Gillette, some Jingdezhen craftsmen try to copy past imperial wares for selling at higher prices. Jingdezhen is not the only pottery-making location with nostalgia for the past.[27] During my fieldwork in Japan, I came into contact with resident potters who continued the traditional of making Nabeshima-style porcelains just as their ancestors had done before them. In the case of Hong Kong, my informants informed me that earlier generations of craftsmen used to imitate traditional Shiwan pottery style but later generations of ceramicists have since turned to making pottery with East-West hybrid designs. These hybrid designs often fuse Western and traditional Shiwan or Chinese influence to produce new styles or products. Third, like Arita, many artisans in Jingdezhen have turned to natural gas kilns. Gas kilns enjoy several advantages over wood-fired kilns, including not releasing carbon emissions into the atmosphere, better temperature controls resulting in less defective pieces due to misfiring, easier kiln management without the need for multiple rotational teams of kilns craftsmen to constantly feed the wood fuels into the kiln for 36-48 hours and piped-in gas is also easily transportable than wooden pieces for firing ceramics. Like their Hong Kong counterparts, if timber firing in kilns is needed, Jingdezhen potters were able to outsource these activities outside the province to their counterparts in Anhui who still operates traditional wood-fired kilns.

I turned to art exhibitions and shows to look at contemporary pieces of Jingdezhen artworks. For "The Art of Contemporary Jingdezhen Ceramics" exhibition, I collected brochures of contemporary artistic Jingdezhen works of art.[28] The exhibition featured lively pastel colors that are delicately applied onto a white bone china porcelain base. They are distinguishable from pre-modern works by the richness of the colors, the thin and delicate (almost translucent) body with uniform dimensions, and animation-like drawings that tend to be more lifelike and realist than amorphous. The strokes are almost felt-tip like with little broad bold application of metallic oxides. A caveat is that the exhibition's works on display do not represent all genres of Jingdezhen's contemporary art. I also did observation work at the following event or facilities (see table 10.4).

I observed some cutting-edge modern art pieces of Jingdezhen-made works in Art Basel Hong Kong 2012, one of the largest art show events in the city. Contemporary artists had taken traditional Jingdezhen porcelain forms like the *tianqiuping* (globular vase with long neck) and turned them into modern art forms with political messages about war of liberation and pictures of jet fighters drawn using traditional cobalt blue and white designs fired against a smooth kaolinite bone china porcelain base.

CONTEMPORARY FIRING TECHNOLOGIES

My informant in Hong Kong who belonged to the fourth generation of ceramics artists in the territory recommended looking at Jingdezhen's studio—The Pottery Work as an example of a site that makes contemporary artistic wares. I was interested in the section of its website which introduced a range of electric kilns for sale orders. It is a useful preview of Jingdezhen artistic studios that prefer to utilize foreign-made kilns for making unique creative works. In total, 25 electric kilns are featured on the web page with its distributor supplying these equipment to would-be interested ceramics artists in Jingdezhen and other parts of China and the prices range from RMB24,155 (KM818 Skutt made in USA) to RMB235,200 (DMT-25A 0.51M3 made in Japan).[29] While different in sizes and pricing, there are some common features amongst these advanced kilns. First, most of them have insulators lining their interior (such as firebricks for the USA-made Skutt models). This makes them more energy efficient. Second, all of them are either the shelf-type kilns with doors or the barrel types that open from the top lid. The use of electricity makes the kiln easy to use with precision controls and they are

Table 10.4. Observation Work

Name of exhibition, event or facility	Types of porcelains displayed	Source/reference
The Art of Contemporary Jingdezhen Ceramics	Contemporary Jingdezhen artistic pieces	The Chinese Civilisation Centre City University of Hong Kong, The Art of Contemporary Jingdezhen Ceramics (HK: City University of Hong Kong The Chinese Civilisation Centre, 2012), unpaginated. (exhibition brochure)
Art Basel at the Hong Kong Convention and Exhibition Centre (HKCEC) on May 23-26, 2013.	Contemporary art with some porcelains featured	NA (based on observation study)

useful for producing limited volumes of porcelain which is usually the case for artistic ceramics wares. The smaller electric kilns are also easily transportable given that the smallest kiln is only 52 kg in weight.

CONTEMPORARY MASS-PRODUCED UTILITARIAN WARES

Besides artistic works of arts, modern ways of production have also enhanced Jingdezhen's ability to mass-produce. With precise temperature controls, use of propane gas kilns and stenciled drawings/sticker designs, Jingdezhen kilns can turn out larger volumes of ceramic products for consumers. I collected some artifacts of different sizes, designs, makes and finishing as some samples of mass-produced Jingdezhen products for closer study. The most common lines of utilitarian wares coming from Jingdezhen today are probably the blue and white dragon series stenciled wares which has been in mass production since the twentieth century. Sometimes, the dragon pattern is replaced by the peony flowers designs. Highly representative of standardized Jingdezhen wares, they have reasonably good quality and finishing without an expensive price-tag.

CONCLUSION

According to the brochure produced by Roger Law and his project group for the Victoria and Albert Museum to highlight Jingdezhen's importance in contemporary porcelain art production, as a creative production center, Jingdezhen is attracting foreign talents to its locale, drawn by the knowledge, skills and enthusiasm in the location for producing ceramics.[30] In terms of knowledge, given that it has been producing porcelains since at least Sung and Ming dynasties, Jingdezhen has accumulated the different styles and designs developed through the different pre-modern Chinese dynasties and re-interpreted in the contemporary period. The knowledge accumulated has also gone through creative destruction, deconstructed and rebuilt over time through alternating phases of political turmoil and golden eras of each dynasty. In terms of its creative and artistic identities, the location negotiates carefully between massive industrial (or proto-industrial) mass production and artistic creative production of unique pieces in individual studios. Finally, the energy and enthusiasm of Jingdezhen artisans and artists can be felt in the diversity of products that they produce—in traditional lines and designs as well as unique pieces featured in international art shows.

NOTES

1. Maris Gillette, "Copying, Counterfeiting, and Capitalism in Contemporary China: Jingdezhen's Porcelain Industry," *Modern China* 36, no. 4 (2010): 369.

2. Geoffrey C. Gunn, *History without Borders: The Making of an Asian World Region, 1000-1800* (Hong Kong: Hong Kong University Press, 2011), 264.

3. Ayers, Impey and Mallet, *Porcelain for Palaces*, 233.

4. Ibid., 234.

5. Art Gallery, The Chinese University of Hong Kong, "5. Bowl Decorated in Doucai Style," *Art Gallery Bulletin* 4, no. 4 (1985): 11.

6. Idemitsu Museum of Arts, *Treasures along the Silk Road: Steppe and Oasis Routes/Sea Route* (Tokyo: Idemitsu Museum of Arts, 2001).

7. Ibid., 109–153.

8. Ibid., 143, 150–51.

9. Hong Kong Museum of Art, *Cruising the Universe: Fantastic Animals in the Arts of China* (Hong Kong: Leisure and Cultural Services Department, 2013).

10. Chinese Antiquities Unit and the Design Unit, Hong Kong Museum of Art, *Gems of Chinese Ceramics from the Hong Kong Museum of Art* (Hong Kong: Leisure and Cultural Services Department, 2012).

11. Stephen Little, "Economic Change in Seventeenth-Century China and Innovations at the Jingdezhen Kilns," *Ars Orientalis* 26 (1996): 48.

12. Art Museum, The Chinese University of Hong Kong (CUHK), *Enlightening Elegance: Imperial Porcelain of the Mid to Late Ming, The Huaihaitang Collection* (Hong Kong: CUHK, 2012).

13. I have received many useful comments from other scholars. One of them pointed out that Arita's big kilns were owned and operated by the Nabeshima-han. "Goyo-tashi" is the term referring to traditional products, e.g., Shogun-ke goyo-tashi, (Name of Han)-goyo-tashi, koshitsu goyo-tashi, etc. Some of these family businesses who manufactured their own products made their family name a brand. The reviewer noted that "goyo-tashi" most probably cannot be equated with what is considered as "private" businesses because they did not really exist in Tokugawa Japan. The same reviewer pointed out that, in a given *han* domain, people were roughly divided into a few categories, but regardless of whether they were peasants, merchants, artisans or samurais, all of them belonged to the *han*. The reviewer noted that a given kiln, for example, could be a money-making business owned by some family, but if a leading member of this family was acknowledged by the *han* domain as lower-ranking samurai, then this business becomes operated by the *han*. The system each *han* had (and there were about 300 han in Tokugawa Japan) were different from one other. They were very difficult to generalize.

14. Michael Dillon, "Jingdezhen as a Ming Industrial Center," *Ming Studies* 6 (1978): 38.

15. Medley, *The Chinese Potter*, 171.

16. Ibid.

17. Little, "Economic Change in Seventeenth-Century China and Innovations at the Jingdezhen Kilns," 49.

18. Nigel Wood, "New Discoveries in Chinese Ceramics," 39, accessed October 29, 2013, http://www.haughton.com/system/files/articles/2010/01/27/102/icf_s_1997_4th.pdf.

19. Titus M. Eliens, "DelftWare WonderWare," *Newsletter* 63 (Spring 2013): 48.

20. Ibid.

21. Gillette, "Copying, Counterfeiting, and Capitalism in Contemporary China," 32. Quoted Rose Kerr and Nigel Wood, *Part 12: Ceramic Technology*, Science and Civilization in China 5 (Cambridge: Cambridge University Press, 2004).

22. Ibid.

23. VOC Arita Listed Company, *The Hague city Art Museum and Nishi Nippon Shimbunsha, Arita Delft China* (Japan: VOC Arita Listed Company, 1993), 23.

24. Gillette, "Copying, Counterfeiting, and Capitalism in Contemporary China," 5-6.

25. Ibid.

26. Ibid., 14.

27. Ibid., 5–6.

28. Chinese Civilisation Centre, City University of Hong Kong, *The Art of Contemporary Jingdezhen Ceramics* (Hong Kong: City University of Hong Kong, The Chinese Civilisation Centre, 2012).

29. Pottery Workshop, "Zheng Yi Trading Company," accessed October 30, 2013, http://www.potteryworkshop.com.cn/trading.asp?183.html.

30. Roger Law et al., *Porcelain City: Jingdezhen* (London: Victoria and Albert Museum, 2012).

Chapter Eleven

Yingge – The Town, Its Developmental History, and Energy Use

According to the signage at the Yingge Ceramics Museum, when the Qing (spelt as "Ching" in English in Taiwan) dynasty authorities established effective control over Taiwan in 1760, waves of Chinese migrants arrived on the island and stimulated cross-straits trade and mainland goods soon overwhelmed the local industries. Because of this, local pottery industries in Taiwan such as those found in Yingge were unable to develop effectively during this period. But ironically, according to the same source, even though mainland Chinese migration may have stifled the growth of local industries, they brought along skills and technology for making pottery to Taiwan. According to the Yingge Ceramics Museum's textual banner titled *Tracing Our Past*, the pre-modern growth of the pottery-making industry in Taiwan benefitted from the successive waves of migration of Han Chinese people from the southern coastal areas during the late Ming and early Qing (spelt Ching in Taiwan) dynasties as well as the post-1949 relocation of the Nationalist government to Taiwan. Amongst the clusters of porcelain-making industries in Taiwan, Yingge quickly rose in prominence, making the transition from manufacturing lower value-added earthenware products to higher value-added items like antique reproductions and artistic wares.

Because of its status as a major ceramics-producing region in Taiwan, Yingge has been given the nickname "Taiwan Jingdezhen"[1] by some local travel guides while other nicknames include "ceramics empire" found in the Yingge Ceramics Museum's textual narrative. Like Jingdezhen, Yingge enjoyed the advantage of close proximity to natural raw materials needed for the pottery-making industry. During the Jiaqing reign of the Qing dynasty

(1796–1820), Yingge's pottery-making industry grew quickly because of the local supply of *danantu* which is ideal clay material for ceramics-making.[2] Such raw materials facilitated Yingge's growth in industrial output because the *danantu* clay was easily shaped, malleable and manipulated to create clay products.[3] According to the Yingge Ceramics Museum's narrative, the region that Yingge is situated in has large forest grounds and its kilns historically utilized acacia, bush woods, rice husks and weed for fuel. In this way, Yingge's advantageous geographical position has some parallels with Jingdezhen. Jingdezhen was located near trees that fueled pre-modern pottery-making in that location and the region also had kaolinite clay that was essential for making white bone china porcelain. Similarly, Arita had a clay stone quarry that fed the porcelain-making activities in that town. In all three case studies, the availability of natural resources was an essential ingredient in the growth of ceramics-making towns. Even though timber-fired kilns have become heritage artifacts in the contemporary period, many ceramic artists that I conversed with fondly highlighted some creative merits of timber-firing in kilns. An advantage of wood firing still relevant today is that it remains the only method available for obtaining a wood ash effect and the process can generate random outcomes. An article in a commercial magazine *The Golden Age* describes the randomness of timber-firing as a kind of natural "firing scars (*huohen*)" and the process of wood-firing as "yin-yang changes (*yin-yang bianhua*)." The two statements are related to the randomness of wood-firing as its imprecise temperature controls can sometimes turn out unexpected glazes. This feature is something that all my artistic potter respondents almost universally raised in my fieldwork. Some of them consider precision-control gas and electric kilns to be too predictable in results.

During the colonial period after Taiwan was ceded to Japan with the conclusion of the Sino-Japanese war 1894–95, Meiji Japan started to introduce western-inspired modernization to Taiwan. Modernity gradually introduced new fuels and technologies such as coal-fired square kilns. The Japanese business community gradually saw Taiwanese consumers as a market for their own businesses and invested in kiln constructions at Miaoli, Nantou and other areas. The colonial state facilitated this process. Yingge imported kiln technology for making tiles from Japan, manufactured a variety of industrial ceramics and re-exported them back to Japan.[4] Historically in the postwar years, just like Arita in the Meiji period, Yingge was a symbol of Taiwan's early industrialization. Modern organizations to mobilize local merchant and commercial interests appeared in Taiwan during the colonial period. Guild-like organizations like Yingge Stone Cooperation was established protect local interests in locally-made products such as pottery and tea in 1917. The colonial superstructure saw the transfer of technology passed from the Europeans to the Japanese and then to the Taiwanese. According to the textual narrative displayed at the Yingge Ceramics Museum, Japan played an instrumental role in the Taiwanese ceramics industry's modernization process by setting up a gas-fired climbing kiln (*noborigama*) in Miaoli and a square gas-fired kiln in Peitou. These kilns were likely experimental ones for facilitating the transfer of skills and knowhow to the Taiwanese, just as the Europeans had done for the Japanese at the early stages of their own modernization process. Other modern technologies transferred from the Japanese to the Taiwanese included electrical equipment, felspar which created synthetic substitutes to kaolinite used in making fine bone china and chemical pigments (e.g., cobalt). Because of the importation of these technologies, Taiwan begins to churn out finer porcelain wares, moving up the value-added chain from the "black bowl" era to manufacturing fine bone china. In the prewar and wartime years, initial wares made by Yingge were crude and rudimentary but they satisfied market demand for simple, affordable and functional utilitarian wares. In a caption titled as "The Black Bowl Episode" in the Yingge Ceramics Museum, Yingge and other local kilns satisfied the demand for ceramic kitchenware at a time of supply shortage during World War II by manufacturing simple, crude but affordable pottery wares for daily household consumption. Without the management knowhow or technical knowledge of making fine bone china porcelain, Yingge and other Taiwanese kilns made "black bowls" which were fired at low temperatures, and were therefore highly porous so they readily absorb oils, gravies and sauces. They were also brittle and cracked easily. They were difficult to wash, making the wares less hygienic and smell bad. But they were affordable and served the minimal functional needs of consumers at a time of extraordinary shortage.

After the Pacific war, Yingge became one of the sites for Taiwanese efforts to re-industrialize their economy. According to the textual narrative at the Yingge Ceramics Museum, the waves of new migrants from the mainland who arrived after the civil war ended in 1949, quickly transformed Taiwan's ceramics industry with their specialized knowhow. The new arrivals knew how to make better utilitarian household wares and their efforts eventually pushed Taiwanese porcelain production beyond pre-WWII levels by 1954. This was the first phase of postwar reindustrialization for Taiwan. The new migrants also brought along technologies, the mindset of scientific rationality and experienced skillsets for use in the ceramics-making industry. With the influx of skills and growing confidence in designs and quality, Yingge ceramics producers displayed their wares at the US international exposition after the 1950s in an attempt to reach new consumer markets as well as establish Yingge's reputation as a premium ceramics-manufacturing center in Taiwan.[5] In the 1960s, Yingge kilns also produced fine reproductions of the imperial wares found in the National Palace Museum for sale to senior-ranking diplomatic staffs based in Taipei.[6]

In the postwar years, Taiwan, formerly an agricultural economy, industrialized in the 1960s and 1970s with light manufacturing industries leading the way, e.g., Yingge's mass production of ceramics. Before making high value-added artistic wares and household decorative implements, Yingge was turning out mostly lower-value added earthenwares and crude industrial products like bricks and tiles. In other words, high value tableware production, reproductions of imperial wares modeled after those in the National Palace Museum as well as artistic ceramics works are relatively newcomers in the town's industrial history. Modern technologies imported from foreign countries facilitated Yingge's industrial production. For example, in 1974, textual narratives from the Yingge Ceramics Museum noted that Italy introduced a hi-tech roller kiln with a heating element suitable for manufacturing industrial tiles with an output of approximately seven hectares' surface area worth of tiles daily (at that time, equivalent to 70% of all tiles manufactured in Taiwan). According to Yingge Ceramics Museum, it takes only half an hour to fire the items fed into the kiln and the fuel source can be gas, heavy oil or electricity.

Yingge depended on coal for its early modernization phase. The explanation text at the Yingge Ceramics Museum indicated that, at the height of Taiwanese coal production, 30,000 tons of coal were extracted annually and transported to storage locations like Yingge (third largest in Taiwan). A railway train model exhibit carrying plastic model coal displayed at the Museum symbolized the age of the arrival of coal-fired kilns where coal is transported to Yingge for powering production activities. However, like the other examples in this book, environmental reasons became a primary factor in shifting Yingge kilns towards the use of cleaner fossil fuels. According to the narrative at the Yingge Ceramics Museum, Hsu Tzu-jan (also spelt as Hsu Zi-ran, the first chairperson of the Taiwanese Ceramics Guild) was a pioneer in turning Yingge away from the use of coal energy in kilns to natural gas use instead because of pollution caused by the former in the town. Hsu's other contributions include introducing postwar Japanese technologies to Yingge's ceramics industry (1968) and persuading his colleagues to make copies and reproductions of imperial wares to satisfy collectors' market for these pieces in 1972. This is an important higher-end sector of the porcelain consumer market that contemporary Jingdezhen is also competing in. Hsu's lobbying activities collectively reflected his importance and standing in the community, not just in energy issues but also economic decisions. Because of Hsu's foresights and the others after him, Yingge entered the phase of manufacturing high-value added items based on scientific rationality and imported technological knowhow, particularly from Japan. My online research indicated that the Hsu family continues to be influential in Yingge affairs. Hsu's son, Scott Hsu, commented on his father's legacy and the impact of energy transitions on Yingge in the town's official media. According to Scott Hsu whose father pioneered the use of natural gas in Yingge, the 200-year old Yingge ceramics industry only prospered after the late 1960s when Taipei City kilns were removed after the government banned the use of coal in the city.[7] This quickly cleaned up the environment in Taipei.[8] However, for Yingge which had no such ban, Hsu was critical of coal use in Yingge:

> While the town's production rose, neither product quality nor the local environment benefited. "There were chimneys everywhere and the sky was always gray," recalls Scott Hsu, China Art's president. "The pollution was so bad that two fifths of the townspeople had tuberculosis."[9]

This description of the polluted environment had correlations with other examples found in this book, Arita and Jingdezhen, both of which had their own similar experiences with managing polluting emissions. Hsu's narrative about upgrading product quality and environmental standards was made possible by the technological factor and cleaner fuels. This resonated with similar phases in the developmental histories of Hong Kong, Arita and Jingdezhen as well.

Besides environmental factors, gas kilns revolutionized Yingge in other ways. According to Scott Hsu, gas-fired kilns are "easier to control and more stable compared with the coal-burning variety and are much more environmentally friendly... improving product quality and minimizing failure rates literally overnight."[10] The Yingge Ceramics Museum's textual narrative identified square kilns and the emissions from its coal-fired combustion chamber as the main reason behind the pollution in Yingge. Natural gas, unlike coal and oil, give off minimum emissions and zero carbon release. Gas made it possible for Yingge workshops like Scott Hsu's family-owned company China Art to make fine imperial ware reproductions starting from 1972 "combined with Japanese raw materials, kilnsmen and glaze painters."[11] Gas-fired kilns had precise temperature controls and user-friendly handling. The production of fine wares gradually pushed Yingge up the value chain in the ceramics industry and it helped to stave off new competitive pressures like the plastic revolution in the postwar years where plastic utensils caused a decline in Yingge tableware sales. With the global recognition of Yingge's artistic works and high-quality ceramics toiletries implements, Yingge created a distinct identity for itself and attracted global and local talents to set up shops in the town. Currently, according to Xinhuanet (Chinese official media), there are over 100 artists who set up their workshops in Yingge.[12] Taiwanese and Hong Kong information sources have higher numbers. According to the Taiwanese authorities' online sources, there are 900 factories and a few hundred retail stores and workshops.[13] These factories and workshops continue to produce five main categories of porcelain products: artistic ceramics, ceramic bricks, architectural and toilet fixtures like toilet bowls and wash basins, utilitarian wares as well as fire-resistant materials like scissors and missile cones.[14]

LITERATURE REVIEW

Richard Stites' article on Yingge is an informative publication on the town's contemporary history.[15] He situates Yingge's development within Taiwan's postwar industrialization, particularly the role played by small and medium sized enterprises (SMEs). Amongst the literatures I reviewed, Richard Stites' 1982 article remains the most informative on contemporary history of Yingge. Stites focuses on "small scale" industries which he defined as "factories using relatively little capital and employing 100 or fewer workers" and Stites quoted Warmelo in arguing that "the predominance of small, individually owned and operated enterprises, established with limited capital and utilizing family labor to the largest extent"[16] is particularly prominent in Taiwan.[17] Stites' article has some similarities with Tolliday and Yonemitsu's[18] case study on Arita. For my literature review section in this chapter, I made extensive use of both writings and examined them in a comparative format. Both articles focus on small companies as the backbone of Yingge and Arita's manufacturing sectors. Stite divides small-scale firms into companies with ten or less personnel and Tolliday and Yonemitsu focus on those hiring between 1-29 individuals.[19]

While Tolliday and Yonemitsu's case study on contemporary Arita examines a region that had been exporting porcelain for centuries, Yingge is a relatively newcomer in the modern mass-manufacturing sector, making mostly lower value-added earthenware like pots, jars, rough-use bowls and toilet fixtures from the early 1950s to the early 1960s.[20] Like Hong Kong's ceramics manufacturing industries located on Peng Chau as well as in the main island in the 1970s and 1980s which declined with the influx of cheaper Chinese-made imports, Yingge was outcompeted by imports when controls were lifted and facilitated the entry of better-quality industrial ceramics into Taiwan. However, Yingge survived by going into higher-value added products, making tablewares, reproductions of antique wares and artistic ceramics for exports and this resulted in an economic boom for the town by the early 1980s.[21] One good example of a product that Yingge manufacturers soon came to dominate in Taiwan was the production of flush toilet fixtures. Some Taiwanese entrepreneurs prospered from this line of products, e.g., the King of ceramics toilet implements, He-Cheng Chiu (1892-1979), was nicknamed as "the emperor of sanitary ware" in the Yingge Ceramics Museum's textual narrative. His manufacturing facility included a 70-meter tunnel kiln facility used to mass produce the white ceramic glaze flush toilet fixtures in large numbers. According to the Yingge Ceramics Museum's explanation, Chiu relocated to Yingge in 1931 and produced the first modern squat toilet fixture in Taiwan coated with a slip glaze. In 1955, his factory manufactured the white ceramic glaze toilet fixtures as well as other utilitarian ware products to replace the old humble earthenware brown bowl.

GATEWAY TO YINGGE

Travelling southwest from Taipei city, I took the train to Yingge for my observation study. Yingge's train station was very similar to its Arita counterpart. Both stations are proximate to storage areas, older shop facades and small-scale retail shops. I would describe both Yingge and Arita train stations' neighborhoods as *shibui* or rustic with a certain small-town charm. In the case of Arita, a nostalgic Showa appearance and, in the case of Yingge, remnant images of Taiwan's booming 1980s greet visitors when they arrive at the train stations of both locations. Not all scholarly works have the same evaluation of the train stations' immediate areas as my observation. Tolliday and Yonemitsu's description of Arita is slightly critical: "The physical appearance and organization of the town are uninspiring."[22] Things have probably improved since Tolliday and Yonemitsu's article written four years ago from the date of this book. During my own observation study to Arita's train station, for example, a tourist booth is installed in the station and there are bicycle rental opportunities for visitors who like to travel around Arita town on a bicycle. Such features are transplantable to Yingge if there are consolidated efforts to build up Yingge's tourism potential and retail sectors that serve visitors. When I was in Taipei and Yingge for this fieldwork, it was not easy to purchase a tourist guidebook for Yingge. Eventually, I found one copy in northern Taipei in Tamkang published by Outdoor Life Books Co. Ltd but it combines the writing on Yingge with three other locations in the peripheries of Taipei City.[23] The narrative in this volume on Yingge tended to focus on its tourist attractions in the peripheries of the town and not exclusively on its pottery-making history and sights. Peripherally, other industries and retailers benefit from the town's main porcelain-making activities. For example, some tea wholesalers can be found alongside the pottery retailers. The train station environment contrasts with the high-tech and new infrastructure feel of Yingge's neighborhood around the New Taipei City Yingge Ceramics Museum marked by newer highways, well-planned industrial areas, gentrified areas and a well-designed museum.

IDENTITY

At the station itself, the town's pottery-making history becomes clearer with display of Taiwan's largest teapot and other large works. By comparison, in the case of Arita, a large dragon design porcelain plate with the words "JR" and displayed on a stand is located at the ticket enquiry section of the train station. They serve the same function of publicly asserting the town's identity as a pottery-making region. At the same time, these large works are also used to show off the technical skills of the craftsmen in the town, given that they are able to produce works that go into the records as the biggest item in various ceramics categories.

In the town's public infrastructure itself, ceramic tiles identifying the town's "cultural corridor (*wenhua yilang*)" is embedded in the concrete pavements. This feature resembles the pavements in Imari City which also has locally-made ceramic tiles embedded in the path as a public display of the town's porcelain-making identity. In the city of Fukuoka (the largest city in the Kyushu region), the terminal station of Hakata also features Arita porcelain tiles integrated into the walls of the train station. Besides ceramic tiles in the pavement in the main street that branches out from the train station (see figure 11.1), Jianshanpu Road, a main pottery-retailing streets in Yingge, is paved with bricks that are fired in the local kilns (during public holidays, cars are out of bounds here). At this street, tourists search for works of arts while others try creating their own pottery works in the shops.[24]

The facade of the oldest street that used to retail ceramics is preserved in Yingge as part of a heritage trail. The street space has been revamped to make it more visitor-friendly, offering nostalgic-looking brick-paved pavements and retailing classically-designed souvenirs while offering workshop sessions where customers can make their own pottery to bring back as keepsakes.

While there are attempts to preserve the traditional facade and features of Taiwanese history in Yingge, local residents have also added their own artistic interpretations to the town by introducing art forms outside local Taiwanese or Chinese traditional cultures. For example, some residents and potters in Yingge made additional efforts to decorate their main entrance with works of art. These works of art have designs based on images inspired by or drawn from other cultures and civilizations. Because they are not necessarily traditional Chinese designs, these art works contribute to the town's metropolitan feel through publicly-displayed artworks created on their own initiatives. There are similarities between Yingge and Arita in this sense because both Arita and Yingge receive multicultural influences from other cultures. Arita has had a long history of indigenizing Western designs, forms and creative influences. Not all residents prefer foreign-looking art works. Some of them continue to display traditionally Yingge-inspired art pieces.

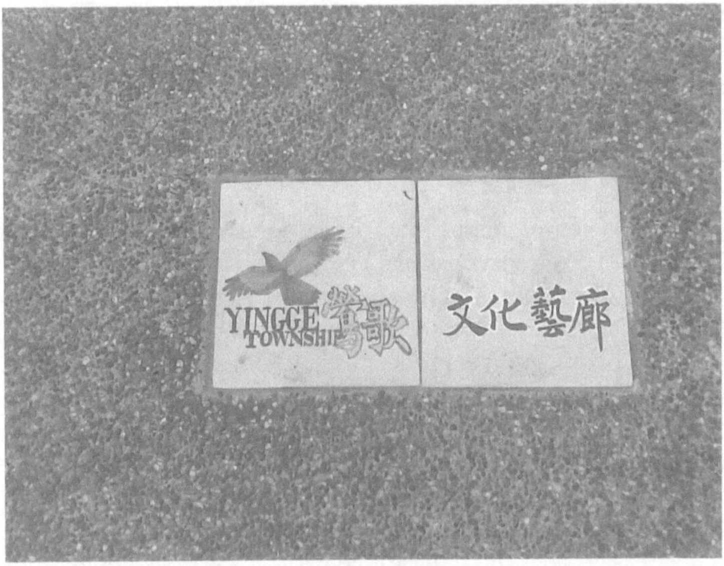

Figure 11.1. Locally-made ceramic tiles are embedded in the concrete pavement of Yingge town with the Chinese words "cultural corridor (*wenhua yilang*)" of the town which refer to its main street. The tiles here feature the picture of an eagle that the town is named after as well as a monochrome red *tianqiuping* and a mud clay teapot made by the town's artisans. The picture is taken during my fieldwork trip to Taiwan in March 2013.

Alongside the multicultural feel in some sections of the town, there are also historical buildings along the main street which have kept their original colonial facade in intact despite the ravages of time. These old residences and shops, if gentrified, have good potential for heritage tourism, a significant and still growing industry in Yingge. Culturally, there are also local practices and customs that have survived the modernization phase of the town's history. Local inhabitants, particularly the elderly, continue to observe customary traditions. These traditions co-exist with the bohemian and artistic atmosphere of the town, giving Yingge a curious mixture of young, hip and Yuppy-ish atmosphere with Zen cafes, contemporary art works, new galleries and older retail outlets, aging neighborhoods and traditional festivals.

SUPPORTING INDUSTRIES

The town is a self-supporting entity with its own town center, retail shops, eateries, grocery suppliers and other features of a thriving local economy. Many of these shops are housed in old shophouses which have narrow five-foot ways but their entrances have just enough space for customers to park their scooters at the entrance. Scooters remain a popular transportation tool in Taiwan. The typical shop houses in downtown Yingge is about two to three storeys tall with balconies on every floor.

DESIGN INNOVATION

Yingge's exposure to international artists and exhibitions facilitate the town's ability to create new designs and forms. I was able to observe local indigenization of traditional Chinese designs used as architectural implements in the public infrastructure of the town. Besides industrial ceramics, according to the Yingge ceramics museum, utilitarian wares, particularly manually-applied blue and pink glazes sealed in a transparent slip glaze have become a representative feature of Yingge table ware. During my fieldwork, I bought some samples of such wares for further study. When I purchased these wares in the main street of Yingge, the shop owner told me that such classically-designed Yingge wares are no longer in production and she considered them as decades-old relics of the past. The wares were thick and had a heavy body and the glaze is similarly thick with visible bubbles that appear during the firing process. The pink colors are typically uneven and form a bulge against the flush surface of the creamy white or green glaze. They do not qualify as fine bone porcelain china with thin bodies but rather resemble stone wares with thick creamy glazes.

According to the New Taipei City Government, Yingge was also responsible for creating the 1950s Carmine Glaze, imitation traditional Blue and White Glaze in the 1960s, and the Crystal Glaze of the 1980s, all of which have been emulated by other kilns in Taiwan.[25] Because of the creative nature of the artistic works produced in the town, the New Taipei City Government came up with the "Yingge Wares Branding Project" in 2008 to create a Yingge Wares trademark (Registry: 01427222) for items certified by Intellectual Property Office, Ministry of Economic Affairs, R.O.C.[26]

When I was at the Yingge Ceramics Museum for observation study, there was an ongoing exhibition on local pottery and the captioned signage for the exhibition reflects the idea of innovation in Yingge:

> This category is aimed to acknowledge utilitarian ceramic works that are innovative in the development and planning of function, design and aesthetics, and integration of the two. They are the embodiment of aesthetics of local life and culture in Taiwan...

Several elements in this Museum caption are noteworthy. First, there is a close connection between ordinary utilitarian tableware and the idea of ceramics innovation in Yingge works. Some of these ceramic works with innovative features are put on display at the Museum. This reflects continuing attempts by Yingge and other Taiwanese potters to produce high-quality table wares with innovative designs for higher value-added customers and avoid competition with lower-end tableware producers. It is the same strategy that Arita producers indicated to me in my fieldwork. Both Yingge and Arita are searching for new functional designs that can attract well-paying customers. In the case of Yingge, indigenously-innovated crystal glazes promoted the town's potteries in the 1980s while new Taiwanese indigenous styles are being promoted in the current exhibition that I attended. Amongst the contemporary works I saw during my observation study, some of the designs on sale at the Yingge Ceramics Museum were based on environmental and multicultural themes. It bears some similarities with my observation field work in Arita which is similarly promoting new hybrid designs based on environmental themes. For example, during my observation study in Koransha, table wares with African safari animal designs were the latest trends in their product designs. Another trend that was visible in the new Taiwanese artistic ceramic designs was the use of a mixed media approach in making ceramics works, e.g., integrating straws into a ceramics work before cutting and reshaping them after baking the porcelain piece in the kiln.

TECHNICAL INFRASTRUCTURE

According to the textual narrative found at the Yingge Ceramics Museum, the rise of the small ceramic studios (as opposed to their large industrial counterparts) started in the 1980s when individualistic styles were valued (similar to Hong Kong's postmodernist consumption patterns) and Yingge attracted creative talents to produce unique works in the town. Yingge provided several important services during this period, including retailing specialized art products and tools, a centralized location for technical skills and knowhow, the presence of kiln choices and the one-stop location of the town's creative workshops. In this sense, Yingge, Arita and Jingdezhen are similar in providing one-stop services for ceramics makers. Travelling through the town for my observation studies, I was able to locate workshops that specialized in *labei* (in hanyu pinyin Romanized spelling) or the technique of hand-throwing ceramic clay. This feature is different from my case study of Arita where most of the shops at the main street were retailers of pottery products rather than ceramics-making workshops. Workshops like this one become the inheritors of traditions and the vanguard of the lost art of hand-thrown manual work in making pottery in Yingge. At one point of Yingge's history, hand-throwing techniques were in danger of dying out, particularly in the economic fast-growth era where molded ceramics were more efficiently produced. Stites' article notes that, as early as 1982, there are only a few potters remaining in Yingge who are still skilled in the technique of *labei*.[27] *Labei* has been gradually replaced in industrial mass production of ceramics by molded porcelains – basically molding two halves of a porcelain ware and then combining them to form a whole piece. In some ways, the manual technique of *labei* or hand-throwing survives because of a DIY (Do-It-Yourself) fad[28] amongst visitors to Yingge who are keen to create and return home with their own ceramics works as a keepsake and souvenir from the trip.

Energy infrastructure is another important production-related feature in Yingge. According to the text displayed at the Yingge Ceramics Museum, natural gas for pottery-making was first introduced into the town in 1971. Compared with other case studies in this book, this was relatively early. Japan, which was the most advanced economy in East Asia at that point of time, completed its transition away from coal energy in the 1960s and, by the 1970s, turned to oil alternatives to prevent overdependence on one single source (especially after the 1973 oil shock). Nuclear power and natural gas were two policy directions in Japan's energy use policies in 1971. As for Jingdezhen, according to Maris Gillette, the transition towards the use of gas kilns took place in 2002 when private enterprises imported gas kilns into the pottery-making town.[29] In Hong Kong, amongst the fourth generation artistic potters that I interviewed, only one of them possess a gas kiln and it was donated to her by an older artist. In utilizing natural gas, the most important infrastructure in Yingge is the pipeline that runs from Miaoli to Yingge,[30] facilitating the town's conversion from coal use to natural gas use, something already full in place by the early 1980s. By the time I viewed contemporary ceramics exhibits at the New Taipei City Yingge Ceramics Museum, a significant number of the artistic works on display used gas kilns and the temperature readings of gas kiln-firing hovered at around an average of 20 to 70 degrees Celsius higher than the works on display which were fired using electric kilns. However, this may be due to a personal choice of fuel selection made by the artist rather than the kiln's capabilities because electric kilns can easily fire up till 1,300 degrees Celsius, although the works that were fired using gas tended to be larger in size as well. Some artistic works were created using electric kilns fired at 1,200 degrees Celsius. The popular use of gas and electric kilns is a similar feature found in my Hong Kong fieldwork where a large majority of the artistic potters preferred the use of electric kiln due to limited industrial workspace in Hong Kong. In almost all the locations covered in this book, respondents and artists indicated that electric kilns were the easiest to operate with their precise controls.

As for the industrial ceramics producers (mainly those Yingge firms that manufacture bricks and tiles), examples of the large industrial kilns they used were featured in the Yingge Ceramics Museum. The following narrative about kilns was drawn from my observation fieldwork conducted at the

Yingge Ceramics Museum. The first industrial kilns that appear in Taiwan were probably the square kilns, some of which were introduced during the colonial period by the Japanese and they ran on either gas or coal fuels. The square chambers facilitated the even distribution of heat, different in principle from the Jingdezhen egg-shaped wood-fired kilns that made use of differential proximity from the heat source to fire wares with different temperature requirements at various corners of the chamber. The coal used for firing the square kilns also makes the preparation time for firing shorter (requires only a few days to prepare) and less manpower-intensive compared with timber-fired kilns. These square kilns had simple square chamber design that could be fired up to 1,250 degrees Celsius and they were eventually replaced by the shuttle kilns discussed below.

Shuttle kilns were gas-powered and, according to the Yingge Ceramics Museum, resulted in greater accuracy in temperature controls for manufacturing imperial ware reproductions and they had two cars moving on tracked rails to enter and exit the kiln that allowed the potters to continue working on the clay items while the kiln remained hot. The heat sources on the sides of the kiln can be adjusted for precise tuning of heat and oxygen input for manufacturing bone white china porcelains. The shuttle kiln is the forefather of the tunnel kiln. The tunnel kiln uses a cart system to introduce clay items into the kiln for preheating, heating and cooling while the shuttle kiln uses only two cars for the same purpose. According to the tourist guidebook that I acquired during my fieldwork, the 25-feet Guzao kiln is the only surviving traditional tunnel kiln left in Yingge.[31] My fieldwork in Arita focused on tableware manufacturers like Koransha and Fukagawa and so no large-scale contraptions like the 70-100 meters tunnel kilns were observed. The only similar mechanized industrial kiln I saw in Arita was a conveyor-belt system kiln where clay items were placed in modestly-sized cargo boxes that held onto the clay items as they enter the kilns for baking. In my fieldwork study of the JCCAC, Fo Tan and Chai Wan studio kilns in Hong Kong which produce mainly artistic works, I could not spot a similar facility (conveyor belt, shuttle or tunnel kilns) in these locations. This indicated Yingge's historical roots in industrial ceramics production which required large tunnel kilns for firing sizable volumes of products in one go to save energy costs. Therefore, the baking chamber of industrial kilns must be large enough to accommodate large numbers of unfired clay items. These operations do not require the delicate temperature controls of fine bone china tableware.

VISITORS AND TOURISTS

The town has one of the world's foremost collections of Taiwanese pottery housed in the New Taipei City Yingge Ceramics Museum (established in the year 2000, please refer to figure 11.2). Chao Chen-Wei and his team's quantitative study indicated that the Yingge Ceramics Museum and the historic ceramics street are the main tourist draws in the town.[32]

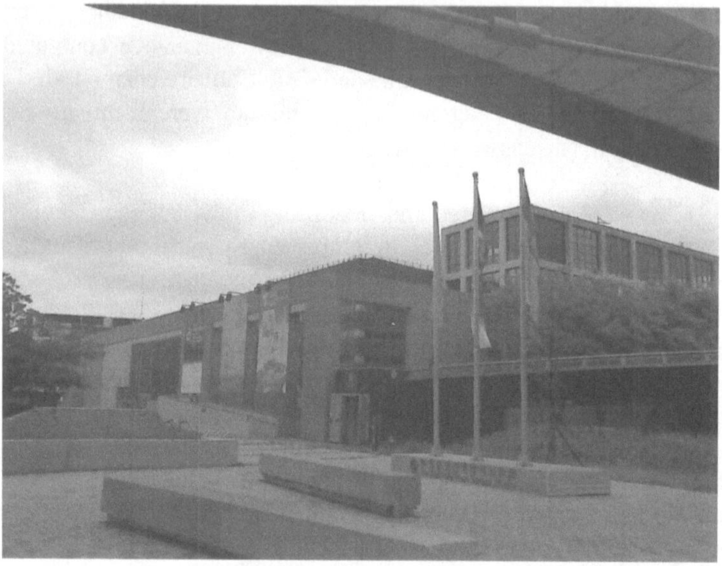

Figure 11.2. A well-designed museum, the New Taipei City Yingge Ceramics Museum, which displays local Taiwanese works of ceramic art sits in the town of Yingge. It has probably one of the world's largest collections of Taiwanese-made pottery. One attractive feature of the museum was that all its exhibits had very detailed information about each individual piece on display, recording information about their glazes, temperatures fired, technique, type of kiln used, etc. Such detailed information were absent in most heritage or museum establishments that I visited in my fieldwork. The photos indicate generous use of glass panes for the building structures. A ceramic installation sits astride the cascading fountain outside the museum. The installation conveys the museum's aspirations to collect or commission works that break the traditional mindset which defines conventional styles, forms and shapes rigidly. Another contemporary design in the form of a ceramics shampoo foam-shaped installation art can also be found in front of the New Taipei City Yingge Ceramics Museum.

Besides the physical building and its exhibits, the museum also organizes heritage activities like timber-firing sessions to ensure the survival of traditional art forms and techniques. Besides the Museum and the historical ceramic street, other tourist attractions in the town include: The Yingge Rock which overlooks a beautiful view, the night food bazaar at Nanya Road and, in addition to natural scenery, there are local delicacies which include meatballs and localized sushi.[33] Like the Arita teacup cafe that I visited during my fieldwork there, Yingge has a similar well-known *nijiang* ("mud-water") porridge dish served in a locally-made lidded container pottery box.[34] In both locations, local cuisines are integrated with local pottery products for sale as souvenirs to visitors and local residents.

CONCLUSION

To summarize the energy development and the history of fuels use in Yingge, I created a chronological table 11.1 for this purpose.

In Table 11.1, I have classified Yingge's energy development into five phases, starting from the importation of traditional skills and knowhow from the mainland, moving on to the modernization phase, indigenous production of utilitarian wares, completion of natural gas infrastructure construction and the innovation of new glazes at Yingge made possible by precision firing. Chronologically, in the pre-modern and wartime periods, these five phases transit from a positionality of traditional skills, limited production and use of traditional fuels; to modernity and mass production made possible by fossil fuels; to a phase of localized demand created by wartime conditions. In the postwar period, newer environmentally-friendly sources of energy such as natural gas are used to create innovative new glazes from the 1950s to the 1980s that defined the town's expertise. Visualized as a timeline, Yingge moved from a position of low levels of technological knowhow to making glazes that are recognized globally. The town has also transited from its status as an importer of light manufactured goods to limited production to meet local demand and finally a producer of high value-added artistic wares for the well-informed and discerning consumer. The last phase probably best reflects Yingge's contemporary transition from a mass manufacturing base (a status that characterized most of its postwar economic catch-up years) to a retail center for ceramics transshipments from mainland China and other areas. As part of this transition to a service industry and in terms of future challenges, Chao Chen-wei and his team's quantitative study are important in pointing out the dilemmatic effects of tourism development in Yingge:

> Tourism development brought more benefits with a creation of employment, increased in revenue, improvement of facility and infrastructure, increased local pride and community spirit, promotion of local culture, and improvement of destination image. Concomitant negative impacts resulted from tourism development also result in environmental damage, life disturbance, traffic congestion, and ecological devastation.[35]

As Yingge shifts into higher-value added artistic ceramics production and gives up more industrial manufacturing sectors (a process that started since

Table 11.1. Historical Evolution of Energy Use in Yingge

Fuel	*Period of Introduction*	*Source*
Use of traditional fuels and the transfer of traditional skills		
Timber-fired kiln knowhow to make simple earthenwares in Yingge arrived with migrants from the mainland. According to the Yingge Ceramics Museum's narrative, the region that Yingge is situated in has large forest grounds and its kilns historically utilized acacia, bush woods, rice husks and weed for fuel.	Late Ming and Qing dynasty	Yingge Ceramics Museum's explanation text titled *Tracing Our Past*
Transfer of modern technologies		
Introduction of gas-fired kilns in Miaoli and Peitou by the Japanese. The square kilns introduced by the Japanese use coal.	The colonial period starting from 1894-5	Textual narrative found in the Yingge Ceramics Museum
Indigenous production		
Taiwanese kilns made "black bowls" which were fired at low temperatures.	During the Pacific War	Textual narrative found in the Yingge Ceramics Museum
Heavy oil was used in kilns to fire industrial products like mosaic tiles	Early postwar years	Textual narrative found in the Yingge Ceramics Museum
Natural gas use		
Hsu Tzu-jan (also spelt as Hsu Zi-ran, first chairperson of the Taiwanese Ceramics Guild) was a pioneer in shifting Yingge away from coal energy to use natural gas instead because of pollution caused by the former in the town.	1960s and 1970s	Jim Hwang, "Going for Broke," Taiwan Info, accessed November 3, 2013, http://taiwaninfo.nat.gov.tw/fp.asp?xItem=24038&CtNode=124.

Fuel	*Period of Introduction*	*Source*
Natural gas for pottery-making was first introduced to the town	1971	Textual narrative at the Yingge Ceramics Museum
Pipeline laid from Miaoli to Yingge, facilitating the town's conversion from coal use to natural gas use	In place by the early 1980s	Richard Stites, "Small-Scale Industry in Yingge, Taiwan," *Modern China* 8, no. 2 (1982): 251.
Development of sophisticated indigenous glazes		
High-fired indigenously produced glazes: 1950s Carmine Glaze, imitation traditional Blue and White Glaze in the 1960s, and the Crystal Glaze of the 1980s	Postwar	New Taipei City Yingge Ceramics Museum, "Yingge Wares – Yingge Ceramics Industry Rejuvenate," accessed November 2, 2013, http://exhibition.ceramics.ntpc.gov.tw/yinggewares/en/index_2.html.

the early 1980s), it will have to rely on the retail sector to support its local economy. The transition from a manufacturing to a service economy mirrors the general economic development of all advanced developing or advanced economies in East Asia. The accompanying benefits are visible from my trip. Local residents display a strong sense of pride as they go about decorating their residences and shops with locally-made Yingge ceramics. Employment in restaurants, cafes and retail outlets occur as manufacturing jobs for earthenware and lower value added industrial ceramics manufacturing declines. Tourism requires an upgrading of public infrastructure and the Yingge Ceramics Museum is a good example of the gentrification process encouraged by the authorities working in tandem with creative workers. The image of Yingge is also enhanced by the international exchanges encouraged by both its private and public sectors. The energy switch to gas completed by the early 1980s resulted in a cleaner environment. All these developments facilitated the town's emergence as a tourist destination. But as Chao Chen-Wei and his team warns, if the tourist and visitor numbers continue to increase, further upgrading of facilities are needed to accommodate increased arrivals. The stakeholders in the Yingge community also need to get together to discuss how to protect their cultural assets and non-tangible resources. Equilibrium needs to be reached between commercial and heritage preservation goals.

NOTES

1. I spotted the use of this nickname in two publications: "Yingge shao" [Yingge Ware], *Golden Age Magazine* (Hong Kong: Golden Age/Senior Citizen Home Safety Association, 2012), 57; Outdoor Life Books Co. Ltd. Tour Information, *Yingge, Sanxia, Ulai, Beiyi Gonglu* (Taiwan: Outdoor Life, 2009), 16.
2. Outdoor Life Books Co. Ltd. Tour Information, Yingge, Sanxia, Ulai, Beiyi Gonglu, 16.
3. "Yingge shao," 57.
4. Ibid., 58.
5. Ibid.
6. Ibid.
7. Jim Hwang, "Going for Broke," Taiwan Info, accessed November 3, 2013, http://taiwaninfo.nat.gov.tw/fp.asp?xItem=24038&CtNode=124.
8. Ibid.
9. Ibid.
10. Ibid.
11. Ibid.
12. Shasha Deng, "Ceramics Workshops in Yingge, Taiwan," *Xinhuanet*, November 2, 2013. http://news.xinhuanet.com/english/photo/2013-06/08/c_132442112_5.htm.
13. This statistic matches that given in a Hong Kong-based publication: "Yingge shao," 57; Hwang, "Going for Broke."
14. "Yingge shao," 58.
15. Richard Stites, "Small-Scale Industry in Yingge, Taiwan," *Modern China* 8, no. 2 (1982): 247-79.
16. W. Van Warmelo, "The Development of Small Enterprises in Taiwan, Republic of China," *Industry of Free China* 28, no. 5 (1967): 4.
17. Stites, "Small-Scale Industry in Yingge, Taiwan," 248.
18. Tolliday and Yonemitsu, "Microfilms and Industrial Districts in Japan," 29–66.
19. Ibid., 30; Stites, "Small-Scale Industry in Yingge," 248.
20. Stites, "Small-Scale Industry in Yingge," 251.
21. Ibid.
22. Tolliday and Yonemitsu, "Microfilms and Industrial Districts in Japan," 44.
23. Outdoor Life Books Co. Ltd. Tour Information, *Yingge, Sanxia, Ulai, Beiyi Gonglu.*
24. Ibid., 16, 20.
25. New Taipei City Yingge Ceramics Museum, "Yingge Wares – Yingge Ceramics Industry Rejuvenate," accessed November 2, 2013, http://exhibition.ceramics.ntpc.gov.tw/yinggewares/en/index_2.html.
26. Ibid.
27. Stites, "Small-Scale Industry in Yingge," 16.
28. Outdoor Life Books Co. Ltd. Tour Information, *Yingge, Sanxia, Ulai, Beiyi Gonglu*, 16.
29. Gillette, "Copying, Counterfeiting, and Capitalism in Contemporary China," 33. Gillette also quoted from this source in discussing about the timber supply industry: Fang, *Jingdezhen minyao*, 179–81.
30. Stites , "Small-Scale Industry in Yingge," 251.
31. Outdoor Life Books Co. Ltd. Tour Information, *Yingge, Sanxia, Ulai, Beiyi Gonglu*, 20.
32. Chen-Wei Chao et al., "The Debate of Environmental Impact and Tourism Development: Perspective from Local Resident and Businessman," *Mediterranean Journal of Social Sciences* 4, no. 10 (2013): 411.
33. Outdoor Life Books Co. Ltd. Tour Information, *Yingge, Sanxia, Ulai, Beiyi Gonglu*, 17.
34. Ibid.,22.
35. Chen-Wei Chao et al., "The Debate of Environmental Impact and Tourism Development," 414.

Chapter Twelve

Conclusion

This writing started off with three research questions in mind: The impact of energy use and the development of kiln technologies in the case studies that I have included in this volume; the second aspect is to examine the mobilization of community resources for ceramics-making activities in the case studies; and the third aspect was to look at how pottery-producing communities distributed their products effectively. Amongst the four case studies that I examined, they can be divided into two main groups when it comes to historicity of the four communities. Jingdezhen and Arita belongs to the category of kiln locations that achieved proto-industrialization and mass production in the pre-modern age. In fact, both communities are world-class producers of ceramics since at least the seventeenth century. The kitchen Qing wares found in my fieldwork in Hong Kong's Lamma Island are a good material artifact example of pre-modern export wares originating from Chinese kilns (including Jingdezhen) that were distributed regionally and globally.

In contrast, in the second group, Hong Kong's Fo Tan and Chai Wan studios as well as Taiwan's Yingge kilns are postwar producers of fine ceramics and artistic wares. Qing dynasty-era (from the reign of Jiaqing onwards) Yingge has a longer history than Hong Kong's studio kilns when it comes to making industrial ceramics, but in the mass production of fine utilitarian wares, they are contemporaries with each other. Both locations started off making crude utilitarian or industrial wares before the cost factor compelled their potters and ceramicists to move up the value chain. Chapter One went into the details about the pre-modern distribution of Jingdezhen and Arita ceramics to the rest of the world in a trading system where the Europeans traders were intensely involved. In this global trading system, export ceramics becomes an important marker of foreign trade between states and economies. The advantage in this inter-regional trade tilted towards the

European producers when they developed their own industrial techniques and synthetic substances to manufacture fine bone china porcelain alongside the ongoing industrial revolution. European consumers became less dependent on East Asian supplies and eventually transferred their technologies to willing learners like Japan. With the ushering-in of the modern period, participation in international expositions became a ritual to initiate ceramics producers into the modern marketplace starting from late nineteenth century. New markets such as New York, Philadelphia and Chicago for Koransha and Noritake porcelain manufacturers were opened up. In other words, the history of the ceramics trade was intimately linked to East Asian industrialization.

In all my case studies, the evolutionary progress of fuel use is a universal commonality. All four case studies experienced the transition from timber to fossil fuels and to low-emissions gas as well as electric kilns. This transition also reflects the direction of the spread of kiln technologies from the Europeans to the Japanese and then the Chinese (in this order), which is the reversal of the proliferation of pre-modern technologies from China to Japan and then to Europe. In the case of Yingge, it was an example of the first modern economy (Japan) in East Asia rather than the Europeans transferring technologies to other Asian countries, signaling one of the first historical instances of technological transfers from an East Asian country to another. In the modern age, the Europeans built experimental kilns for the Japanese while the Japanese state and private sector left behind some coal-fired kilns in Baoshan taken over by the Chinese after World War II. The Japanese also invested in colonial-era kilns in Taiwan which were similarly inherited by the postwar Taiwanese authorities. The product types made by increasingly advanced pottery kilns and energy fuels followed a chronological progression, from crude earthenwares and industrial products to standardized high-quality fine bone china porcelain wares. Efficiency (measured in terms of reduction in numbers of defect) and productivity (the number of items that can be baked in one single firing and the amount of manpower needed for each firing) also increased with more efficient fuels.

In many ways, the fuels transition in kilns resembled the developmental histories of the local economies of ceramics-manufacturing towns. Yingge is a good example of how a pottery-making town entered modernity when the colonial government invested in new kiln infrastructure in Taiwan. Driven by the promise of a new consumer market and also the draw of a cheap labor force for production, Yingge became a beneficiary of the transfer of technology that kick-started its prewar industrialization process. In the postwar years, the conversion of the town's kiln into gas-powered ones became the next stage of industrialization when Yingge was able to produce fine porcelain table wares and antique reproduction wares for sale on the international market. Its aritisans' expertise moved beyond merely producing crude industrial earthenwares. Environmentally, along with this fuels progression, the amount of emissions released into the atmosphere also decreased correspondingly. But not all of it was unequivocally good news for potters. Advancements in productivity and firing effectiveness were mitigated by over-predictable firing results that hampered creative processes. With the advent of gas and electric kilns, because temperature controls had become so precise, it was no longer easy to achieve surprises in glaze outcomes. Glaze colors had become so predictable that some ceramics artists are unexcited about the outcome.

The efficiency of the fuels also triggered off reactions related to the artists' creative process and consumer preferences. A segment of both communities yearned for individually-made ceramics products that were not mass-produced or standardized. In connection with this, I examined the tensions brought about by ideas related to postmodernism in the tussle between unbridled capitalism and postmodern impulses to consume unique handmade items ethically. The abstraction of postmodernism opens itself up to subjective interpretation and this is manifested in the different ideas about folk wares, traditional crafts, nostalgic production and studio works. The commonality amongst these ideas is the loosely united front of resistance towards standardized consumption and uniformity of tastes. In this way, the democratization of production means to the individual potters made possible by new kiln technologies offer an alternative to mass production. Some ceramicists have gone beyond ethical consumption and individual production to become activists to protect traditional ways of crafts-making in areas that are literally coined as the global system's "last frontiers" (such as Myanmar, also known as Burma).

There are three major reasons for fuels choice in the ceramics-making industry: availability of fuel, productivity and environmental friendliness. In the pre-modern era, availability of fuel was a primary consideration in fuel selection for use in the kilns. For example, Jingdezhen was located near forests while Yingge was located near timber, cereals and other natural organic combustible materials which were used in kiln-firing. Availability of natural resources was a primary factor in the growth of pottery-making town as timber suppliers became an important part of the local economy. The second factor of productivity was an important guiding force in kiln development in the age of protoindustrialization and mass production in the modern age. Export wares needed to be mass-produced to satisfy market demand domestically and overseas. The development of mass production technologies translated to leaner production when the number of defects decreased. At the height of Ming-Qing fascination with imperial wares, fired pieces were destroyed for minor defects in order to keep the perfect pieces for use by the imperial household. As kiln-firing technologies improved, productivity increased and the number of defects declined. Electric and gas kilns with their precise temperature controls eliminated any need for mass destruction of defective pieces (like the pre-modern technique of making imperial wares) to save the good ones. This saved resources, energy and time for the ceramics producers. Insulation typically found in electric and gas kilns retain heat within the chambers, making the equipment more energy-efficient.

Increasing productivity was also a response to competitive cost pressures that in turn determined the location of production bases and the types of wares they make. A good contemporary example is how Hong Kong kilns lost their mass production orders when Chinese imports became available for Hong Kong consumers and transshipments to third countries. Therefore, Hong Kong potters focused on higher value-added artistic ceramic wares and worked in lower-cost and smaller spaces that necessitated the use of smaller electric and gas kilns. In line with a densely-populated land-scarce urban environment, gas and electricity were easier to transmit by piping-in and this further saved transportation costs (by contrast coal needs trucks and railway facilities for transportation). The third consideration of care for the environment came later than the other two. It symbolized a rethink on unmitigated emphasis on quantitative expansion. Arita, Jingdezhen and Yingge all faced a polluting phase at one point of their developmental histories. My informants in Hong Kong showed the strongest consciousness in thinking about the environment when creating their works. Attention was paid to toxicity of materials, recycling processes, reduction of emissions and other forms of rationalized production to minimize wastage. Informants in the other case studies drew their artistic and creative inspirations from natural sceneries and objects found in nature. The concept of the environment, therefore, was integrated into the works in both production as well as artistic processes.

In addition to functional reasons, human factors and subjective emotive factors like aesthetic appreciation also shaped the selection of kiln technologies. Electric and gas kilns are preferred today in all my case studies because of their ergonomic handling, simple operational features and outcome predictability. These are human factors that persuade potters, artists and ceramicists to use fuels that give them the best control over production needs. Emotionally, the use of gas and electric kilns that are easy to use and compact to keep has a democratizing effect on both users and ceramics consumers alike. Energy provision is compartmentalized for the individual user who can now make their own pottery from the comforts of their own home or studio without the presence of a team of kiln hands, suppliers to chop timber or a large space with sloping terrain to construct a traditional kiln. The simplification of fuel access through piped-in convenience (electricity or gas) place production factors into the hands those who desire them.

This feature is facilitated by the option of purchasing raw materials for firing ceramics through online ordering. Using the internet, ceramics-making enthusiasts can now purchase the raw materials like clay and other ingredients they need like glazes through online means. Potters who use these home kilns can now create the designs that they prefer. This also facilitates the production of ceramic goods that are not mass-manufactured and custom-made to the individualized taste. The accent is on variety and the availability of choice in shopping for ceramic products. Sometimes, the availability of choice is predicated on nostalgia, the desire for simpler, more rustic-looking and functional ceramics products made in the past with traditional techniques before the advent of mechanical mass production. The folk pottery movement examined in the preceding chapters is also a consumer reaction to excessive mass production of ceramics and the absence of choice in standardized wares. In the case of the ethically-conscious Hong Kong potters, the use of local materials (such as locally-available minerals to create glazes) and drawing inspirations from local sceneries and landscapes are all part of the local grassroots reactions to over-commercialized mass production systems.

Kiln technologies have always been situated between the contesting priorities of more efficient production and greater expression of individual creativity. Sometimes, the tussles between these two goals are determined by the availability of technologies to the potters and ceramicists of various historical periods from the pre-modern to the contemporary eras. Table 12.1 summarizes the discussions in the preceding chapters on kiln technologies.

The emergence of more energy-efficient and manageable kilns means that traditional kilns are now relegated to heritage purposes. They are preserved to teach younger generations of craftsmen skills in operating a timber-fired kiln. In the case of Arita, large ceramics-producing companies like Fukagawa make it an annual affair to operate the company's only timber-fired kiln in Arita, not only for heritage purposes but also community identification and bonding. Very often, the traditional kilns themselves are historical relics that form an integral and important part of a location's developmental history. It provides residents in that area with a historical awareness of how early industrialization occurred in the local history of the community and how it contributed to the national economy in early stages of development. The kilns and the narratives they represent becomes part of the fabric of community-building (and in some cases of famous kilns, nation-building) and form collective memories that are stored in local and national museums, archives and other institutions that retain such memories.

One tangible economic advantage of heritage preservation of older kilns is facilitating tourism. For heritage tourism to work, the owners of the kiln need to work closely with local residents, the private sector, local authorities, creative workers, heritage agencies and charitable donors to bring about sustainable tourism for the local community that can reach an equilibrium between commercial needs and heritage goals. Some of these heritage sites are remnants from the past (e.g., the Wun Yiu in my Hong Kong case study) while others are reconstructed by heritage enthusiasts (e.g., the *anagama* in my Izumi city case study). These kilns become tourist attractions for heritage

Table 12.1. Discussions in the Preceding Chapters on Kiln Technologies

Type of kiln	Traditional or contemporary	Advantages of this kind of kiln	Its drawbacks
Timber-fired kilns (examples of wood-fired kilns below)	Traditional	Sometimes produces unpredictable results that can turn out to be quite pleasing aesthetically.	Large size, difficult to fit into a contemporary apartment. Illegal to use timber-fired kilns in certain locations. Need the interior of a wood-fired kiln to cool down before entering to retrieve works.
Anagama (cave) kiln	Traditional	Easy to construct with primitive materials	Comparatively less efficient heat retention and difficulties in temperature control.
Egg-shaped kiln	Traditional to modern (late Ming dynasty to the 1950s)	Able to distribute unevenly to fire different wares in one firing. This is useful for firing different wares with differential firing temperature needs.	Same challenges as all wood-fired kilns.
Chinese dragon kiln	Traditional	Can fire with evenly-distributed temperature. Multi-chambered, can fire large quantities of ceramics.	Manpower intensive, relatively imprecise temperature control compared with contemporary kilns
Japanese *noborigama* (climbing) kiln	Traditional	Can manipulate draughts and air input and have varying intensity of heat. Multi-chambered, can fire large quantities of ceramics.	Like other timber-fired kilns, manpower intensive and more imprecise temperature controls compared with contemporary kilns. Different heat intensities at different corners of the chamber.

Jingdezhen dome-shaped kiln	Pre-modern	Advanced during its era.	Manpower needed to manage air input and control the amount of wood fed into the kiln. Need human inputs for wood fuel insertion. Long time to prepare for firing. Firings result in defects as well.
Square kiln	Early modern	Coal fuel more efficient than timber. More even distribution of heat and temperature than the wood-fired kilns. Larger numbers of clay items can be baked in one single firing. Early industrial technologies make it possible to produce large numbers of standardized utilitarian wares for exports. Takes only a few days to prepare for firing.	Pollution.
Shuttle kiln	Modern	These kilns have cars on tracks transporting the clay products for firing. Allow kiln operators to work on the pieces while it is still hot.	Same issues faced as other kilns running on solid and liquid-based fossil fuels.
Tunnel kiln	Modern.	Can fire large number of pieces in one session. An improved version of the shuttle kiln that uses a cart system for transporting clay items for the firing process.	Same issues faced as other kilns running on solid and liquid-based fossil fuels. Occupies a large space.
Beehive kiln	Modern	Uses coal to fire wares more efficiently. Some artists say it has a natural-looking glaze.	Pollution, ease of use lower compared with contemporary electric or gas kiln. Emits carbon and soot.

Coal-fired home kiln	Contemporary	Small enough to fit into the garage or the lawn of a home. If smokeless coal or charcoal is used, then emissions are minimized. Holes by the side for air introduction and wind inputs.	Smokeless coal is expensive. Can only fire small quantities.
Kerosene-fired kiln	Modern	Oil is more efficient than coal for kiln-firing. Some artists say it produces a natural-looking glaze.	Gives off smell and also noise pollution from its fan. Some danger of accidental fires.
Gas kilns	Contemporary	Environmentally-friendly. Little emissions. Gives fired ceramics a natural look like other fossil fuels. Heat retention properties of the body also make it more energy-efficient.	Noisier than electric kilns. Low manpower needs. Small ones can be a one-man operation (OMO).
Electric kiln	Contemporary	Silent and environmentally friendly. Small ones are mobile and portable. Ergonomic and precise temperature control. Ideal for hobbyists, amateurs and recreational potters. Some of my informants have their own home-build filtration systems. Low manpower needs. Small ones can be one-man operation (OMO) --- e.g. the 100V electric kiln types. Heat retention properties of the body also make it more energy-efficient.	Electricity bills can be expensive if a large quantity of ceramics is fired.
Oil drum kiln	Contemporary	Can be home-made. Small fits snugly into an apartment room. Insulation inside the oil drum is heat-retentive.	Can only fire a limited number of ceramics per session.

visitors. Other tourists interested in putting their hands to work when visiting the ceramics-making towns attend workshops to hand-make their own pottery pieces. For the less adventurous, there are retail shops selling souvenirs for visitors bring back as keepsakes. Heritage tourism can help regain employment in ceramics-making towns whose competitiveness has declined in certain industrial activities (like making industrial ceramics for example) due to cheaper alternatives in locations with lower labor costs.

Both Hong Kong and Yingge case studies are good examples of how their ceramics products have lost their competitiveness to cheaper products made in China. Tourism and its accompanying service industries can also help to provide employment for supporting industries (timber suppliers, wood cutters, etc) in pottery-producing towns like Yingge, Arita and Jingdezhen that have been eliminated through the use of newer types of fuels like natural gas and electricity. Yingge and Hong Kong's Wun Yiu authorities and management need to work closely with other community stakeholders to discuss how best they can protect the heritage resources by balancing between heritage preservation and attracting tourist revenues. A balance also needs to be struck between continuing with traditional designs and systems of production to ensure their survival and continuity versus continuous innovation to create new designs that appeal to contemporary consumers. Heritage preservation does not only refer to protection of old historical sites but also the enhancement of a town's pottery-making identity by integrating locally-made porcelain products into the public infrastructure (e.g., porcelain integrated into public grounds, or vases displayed on public streets, etc). Other local cultures are integrated with the pottery-making activities to create unique products for local visitors and tourists alike, e.g., serving local culinary cuisines in domestically-made traditional potteries such as those found in Arita and Yingge.

Amidst the changes that have taken place in the industry, some features remain constant. For example, blue and white fine porcelain remain as evergreen favorites amongst ceramics collectors and consumers. When natural cobalt materials were either low in supply or not available, the European ceramicists came up with alternatives to create the metallic oxide ink for manufacturing blue and white porcelains and used feldspar to make up for the unavailability of kaolinite in Europe. In this way, synthetic materials made blue and white porcelain wares manufacturing possible even without the natural ingredients such as kaolinite. Such technologies were transferred to Japan and then the rest of East Asia in the late nineteenth and early twentieth centuries. Arita and Jingdezhen continue to make traditional blue and whites in three important categories of products. The first category of porcelains produced is new product designs — innovative and creative new blue and white designs to fit contemporary consumer tastes. They can be in the form of functional designs (e.g., hygienic porcelain toilet fixtures with blue and white designs) or aesthetically unique forms (e.g., egg shell thin ceramics with blue and white designs) to appeal to different customers. Functional designs are sometimes shaped by traditions, for example, the ritualistic rules governing the dimensions of Japanese tea bowls. In other instances, they are influenced by ergonomics, the ease in handling various types of pottery wares for different uses in the kitchen, on the dinner table or in the bathroom.

The second category of porcelains is the reproductions. These are wares made to resemble the imperial wares at the National Palace Museums. First made by Yingge potters who use traditional methods to fire them, this sector of porcelains single-handedly revived the Yingge industry economically when its competitiveness in other products like crude industrial earthen wares had declined. Later on, Jingdezhen got into this industry and, when the Jingdezhen timber-fired kilns were shut down in favor of gas kilns, Jingdezhen potters specializing in such wares moved their production sites to Anhui. The third category of blue and whites are the hybrid porcelains—the combination of East and Western-inspired designs. They can be in the form of traditional Japanese Arita designs found on Western-style teacups or Jingdezhen-inspired designs that were copied and indigenized by the likes of Delft and Meissen and then re-interpreted by Jingdezhen for a contemporary audience. Trade facilitated this cross-pollination of design ideas between Europe and East Asia, resulting in forms and designs that incorporated aesthetic features from both sides. Very often, adaptation of hybrid features caters to the consumers' tastes for something exotic and yet still retains some familiar features. These designs are sometimes known as hybrid designs that incorporate East Asian and Western designs in form and aesthetic presentation. Koransha and Fukagawa both produced such hybrid designs which I came across during my fieldwork in Arita. The works made by the fourth generation of Hong Kong potters based on their educational background and East-West influences are another good example of such hybrid art-forms.

The second feature of continuity is that Arita and Jingdezhen continue to be one-stop center for potters and ceramicists alike with the accumulation of skillsets, historical experience, kiln facilities, customer base and a culture of collaboration. The ecology of Arita and Jingdezhen production bases facilitates artists who want to enjoy an atmosphere that favor collaboration and exchange of ideas and opportunities. These factors are unchanged regardless of the different stages of modernization. In contrast, Hong Kong's clusters of individual ceramics studios in Fo Tan and Chai Wan may not offer the same comprehensive one-stop service for all stages of firing. Fo Tan and Chai Wan form another kind of production ecological network consisting of disparate artist studios that help each other informally based on friendship. Fo Tan and Chai Wan are the creations of a chaotic network of individual artists and ceramicists who have project-basis collaborations in a dense urbanized environment where *guanxi* connections can mean getting skilled expertise for free (e.g., photograph for brochures), commissioned works at hotel, or simply networking at social gatherings for exchange of ideas.

There are some attempts in Hong Kong to institutionalize the community network, e.g., the Jockey Club Creative Arts Center (JCCAC) which houses artists (using different media for their works) in multi-storey industrial buildings, is touted as Hong Kong's first arts village. This arrangement is different from Arita and Jingdezhen in the following ways. The JCCAC is not mainly a pottery-making location like the other two. While JCCAC is a gentrified facility designed to mitigate lower tenancy and revamp old industrial sites, by contrast, Jingdezhen and Arita has a long historical tradition in the accumulation of skills and talents for firing ceramics. The legitimacy of a long tradition by itself is an attraction for artists and ceramicists to work in the two pottery-making towns. The presence of kiln facilities in the town is a permanent infrastructure where potters and ceramicists alike can meet to collaborate on projects together. JCCAC is also unable to accommodate large traditional kilns like *noborigama*s and dragon kilns. It also does not have the same natural resources like indigenous clay materials, timber, organic fuels, etc. found in Arita and Jingdezhen. The third feature of continuity is found in the sustained production of classical designs. Reproductions of Arita wares, for example, are not only found in Arita but also in Thailand and China. The latter two production bases manufacture Arita-like porcelain for export to foreign as well as Japan's domestic markets. Ironically, the proliferation of classical Imari or Arita designs to other production bases in the world will ensure the survival of these designs, if they are ever discontinued in the original place of manufacture.

Finally, I have sent this work to many scholars and practitioners for their opinions and I received many helpful suggestions. I have implemented most of them in the main text. Some offered me future directions for a follow-up volume. I will discuss them here as an indicator of possible future writings to come. It comes as no surprise that the group of informants in my project that is most connected with environmental issues continues to exchange views about environmentally-friendly kiln technologies with me. I am grateful for this exchange. Amongst the articles that I received, here are some interesting new developments related to the emergence of new environmentally-friendly kiln models. The methane kiln makes use of methane gas generated by landfill wastes and then transmits them via pipelines to the kiln site.[1] Because it makes use of wastes from the landfills, the cost of energy is cheap and it prevents methane from escaping into the air causing environmental pollution, but the only drawback is that it fires mainly industrial products and stonewares but not fine porcelain.[2] The pallet kiln makes use of unwanted wooden packing pallets by burning them in a large chamber that can consume an entire wooden pallet with nails which drops off harmlessly into the deep chamber without harming the fired clay.[3] The kiln can consume up till 120 pallets each firing[4] and this saves the environment from pallets discarded improperly and also produces tangible output of ceramics products from the burning. These kilns as well as new technologies that I encountered in my fieldwork like the smartphone-controlled kilns where firing can now be activated remotely represent the next generation of environmentally-friendly, energy-efficient, ergonomic and non-manpower-intensive options available for the interested potter.

NOTES

1. William Baker and Joy Tanner, "Two Kilns that Make Smart Use of Waste," accessed October 25, 2013, http://ceramicartsdaily.org/ceramic-studio-equipment/ceramic-kilns/two-kilns-that-help-reduce-environmental-impact-of-firing/sustainable_620.
2. Ibid.
3. Ibid.
4. Ibid.

Bibliography

Ameba Japan. "Yoshida-ya Shōyu." February 5, 2011. Accessed February 15, 2013. http://ameblo.jp/yoshidaya-soysauce/.

Antiquities and Monuments Office (AMO), Leisure and Cultural Services Department. "Pottery Kilns at Wun Yiu Village in Tai Po." Accessed October 21, 2012. http://www.amo.gov.hk/en/monuments_20.php.

Arita Porcelain Lab. "What is Arita?" Accessed March 16, 2012. http://www.aritaware.com/global/.

Arita VOC, Hāgu City Art Museum and Nishi Nippon Shinbunsha. *Tōji no tōzai kōryū ten: Arita, Derufuto, Chūgoku no sōgo eikyō* [East-west exchange of ceramics: Mutual influence between Arita, Delft and China]. Saga-ken, Nishimatsuura-gun, Arita-machi: Arita Vī Ō Shī, 1993.

Art Exploration Society. "Hong Kong Creative Art Explorers Series: Understanding Burmese Lacquerware." May 11, 2012. Accessed September 20, 2012. http://www.meetup.com/Art-Exploration-Society-Hong-Kong/events/62875472/.

Art Gallery, The Chinese University of Hong Kong. "5. Bowl Decorated in Doucai Style." *Art Gallery Bulletin* 4, no. 4 (1985): 11.

Art Gallery of NSW. "Hasami Ware (Japan): Dish with Incised Waves and Lotus Design." May 2000. http://www.artgallery.nsw.gov.au/work/131.2000/.

Art Museum, The Chinese University of Hong Kong (CUHK). *Enlightening Elegance: Imperial Porcelain of the Mid to Late Ming, The Huaihaitang Collection*. Hong Kong: CUHK, 2012.

———. *Filling the Interregnum Ming Mid 15th Century Ceramics from Jingdezhen*. Hong Kong: CUHK, 2012.

Artmap. *2012 Artmap*. Hong Kong: Artmap, 2012.

Association for the Promotion of Traditional Craft Industries. "Amakusa Pottery and Porcelain." Accessed August 1, 2013. http://kougeihin.jp/en/crafts/introduction/pottery-and-porcelain/2867?m=pd.

Ayers, John, Oliver Impey, and J. V. G. Mallet (with contributions by Anthony du Boulay and Lawrence Smith). *Porcelain for Palaces: The Fashion for Japan in Europe, 1650–1750*. London: Oriental Ceramic Society, 1990.

Baker, William, and Joy Tanner. "Two Kilns that Make Smart Use of Waste." Accessed October 25, 2013. http://ceramicartsdaily.org/ceramic-studio-equipment/ceramic-kilns/two-kilns-that-help-reduce-environmental-impact-of-firing/sustainable_620.

Benskin, Elizabeth. *The Arts of Japan: A Teacher's Guide*. Washington: Smithsonian Freer Gallery of Art and Arthur M. Sackler Gallery, 2005.

Bentley, Jerry H. "Hemispheric Integration, 500-1500 C.E." *Journal of World History* 9, no. 2 (Fall 1998): 237–54.

Chan, Michelle. "Last of the Porcelain Painters." *Standard*, January 20, 2007. Accessed September 5, 2012. http://www.thestandard.com.hk/news_detail.asp?pp_cat=30&art_id=36430&sid=11773593&con_type=3.

Chao, Chen-Wei, Oliver F. Shyr, Ching-Sung Lee, Chen-Huang Chao, Tsai Li, and Chih-Sheng Kang. "The Debate of Environmental Impact and Tourism Development: Perspective from Local Resident and Businessman." *Mediterranean Journal of Social Sciences* 4, no. 10 (2013): 410–414.

Chen, Longgen. "Exquisite Ceramics and Calligraphy on Display at CityU Gallery." Accessed October 30, 2012. http://www6.cityu.edu.hk/puo/CityUMember/Story/Story.aspx?id=20070607164825.

Cheung, Sidney C.H. "Remembering through Space: The Politics of Heritage in Hong Kong." *International Journal of Heritage Studies* 9, no. 1 (2003): 7–26.

Chinese Antiquities Unit and the Design Unit, Hong Kong Museum of Art. *Gems of Chinese Ceramics from the Hong Kong Museum of Art*. Hong Kong: Leisure and Cultural Services Department, 2012.

Chinese Civilisation Centre, City University of Hong Kong. *The Art of Contemporary Jingdezhen Ceramics*. Hong Kong: City University of Hong Kong, The Chinese Civilisation Centre, 2012.

Chiu, Ping Kwan. "A Unique Experience of Setting Up and Running the HK First Artist Village." Powerpoint presentation at the Half Day Symposium on Revitalisation of Factory and Old Buildings, Hong Kong Institute of Facility Management, January 22, 2010. Accessed December 13, 2012. http://www.hkifm.org.hk/2010_doc/ppt_20100122_pkchiu.pdf.

Chu, Lung-hsing. "The Meeting of China, Japan and Holland: Dutchmen in Japanese Prints during the Edo Period." Paper presented at the Design History Society Annual Conference, University College Falmouth, UK, September 3–6, 2008.http://rd.cycu.edu.tw/RD2008/H2500/rtd/project6/97%5C26%E6%9C%B1%E9%BE%8D%E8%88%88ok%5CPaper003_The%20meeting%20of%20China%20Japan%20and%20Holland.pdf.

Citistore. "Rishi Taoci Canjuzhan" [Japanese-style porcelain and tableware exhibition]. *Headline Daily*, October 24, 2013, 35.

City Art Museum of Saint Louis. *200 years of Japanese Porcelain*. Introduction and catalogue by Richard S. Cleveland. Saint Louis: City Art Museum of Saint Louis, 1970.

Clark, Hugh. "The Religious Culture of Southern Fujian, 750–1450: Preliminary Reflections on Contacts across a Maritime Frontier." *Asia Major* 19, no. 1/2 (2006): 211–40.

Commissioner for Heritage's Office, Development Bureau. *A Newsletter on Revitalisation: Revitalisation @Heritage*, no. 25. Hong Kong: Commissioner for Heritage's Office, Development Bureau, July 2002.

Commissioner for Heritage's Office, Development Bureau. *Heritage Fiesta 2012*. Hong Kong: Commissioner for Heritage's Office Development Bureau, 2012.

Commissioner for Heritage's Office and AMO. "Tai Po Wun Yiu Kiln Site." Accessed October 16, 2012. http://www.heritage.gov.hk/taipowunyiu/en/background.html.

Committee of Dual Plasticity Ceramics Exhibition. *Dual Plasticity Ceramics Exhibition*, Jockey Club Creative Arts Centre, October 22, 2010–November 6, 2010. Hong Kong: Committee of Dual Plasticity Ceramics Exhibition, 2010.

Currier, Jennifer. "Art and Power in the New China: An Exploration of Beijing's 798 District and Its Implications for Contemporary Urbanism." *Town Planning Review* 79, no. 2-3 (2008): 237–65.

Dargin, Justin, and Tai Wei Lim. *Energy, Trade and Finance in Asia: A Political and Economic Analysis*. London: Pickering & Chatto, 2012.

de Vries, Jan. "The Limits of Globalization in the Early Modern World." *The Semi-Daily Journal of Economist J. Bradford DeLong* (October 17, 2007).

de Wall, Edmund. "Lucie Rie: Modernist Potter by Emmanuel Cooper." *Guardian*, May 25, 2012. Accessed October 20, 2012. http://www.guardian.co.uk/books/2012/may/25/lucie-rie-modernist-potter-emmanuel-cooper.

Deli Communications. *Gourmet Issue* 40 and 41. Hong Kong: Deli Communications, October and November 2012.

Deng, Shasha. "Ceramics Workshops in Yingge, Taiwan." *Xinhuanet*, November 2, 2013. http://news.xinhuanet.com/english/photo/2013-06/08/c_132442112_5.htm.

Department for Culture, Media and Sport (DCMS). "Last Chance to Keep 'World Class' Japanese Porcelain from Princess Diana's Childhood Home." February 2011. http://www.culture.gov.uk/news/news_stories/7779.aspx.

Dillon, Michael. "Jingdezhen as a Ming Industrial Center." *Ming Studies* 6 (1978): 37–44.

Doolan, Paul. "The Dutch in Japan." *History Today* (April 2000): 36–42.

Dreamstime. "Royalty Free Stock Photos: Ruin- Pagan, Burma (Myanmar)." Accessed September 19, 2012. http://www.dreamstime.com/royalty-free-stock-photos-ruin-pagan-burma-(myanmar)-image9838658#.

Eliens, Titus M. "DelftWare WonderWare." *Newsletter* 63 (Spring 2013).

Ercilasun, Konuralp. "Silk Road as a Sub-global Region: A Sphere Emerging from the Interaction of Cultural and Economic Fields." Paper presented at an International Seminar on "Reviving the Silk Route: New Initiatives and Engagements for the 21st Century," International Centre, Goa, Dona Paula, India, February 9–10, 2007. http://www.centralasia-southcaucasus.com/docs/Silk%20Route/Paper_Konulrap_Ercilasun.doc.

Espir, Helen. *European Decoration on Oriental Porcelain, 1700-1830*. London: Jorge Welsh books, 2005.

Evans, Graeme. "Measure for Measure: Evaluating the Evidence of Culture's Contribution to Regeneration." *Urban Studies*, 42, no. 5-6 (2005): 959–83.

Fang, Lili. *Jingdezhen minyao* [Jingdezhen's folk potteries]. Beijing: Renmin meishu chubanshe, 2002.

Flagstaff House Museum of Tea Ware. *Museum Guide*. Hong Kong: Flagstaff House Museum of Tea Ware, n.d. (museum brochure obtained in 2012).

Florida, Richard. *Cities and the Creative Class*. New York: Routledge, 2005.

Foundation of the Shigaraki Ceramic Cultural Park, ed. *Meiji no yakimono: bunmei kaika no yakimono, Ō-bei o fūbishita japonizumu e* [Japonisme for the Western world: Pottery of the Meiji era]. Kyōtō: The Foundation of the Shigaraki Ceramic Cultural Park, 1996.

Fukagawa, Tadashi. *Koimari no bi to roman* [Old-Imari's beauty and romance]. Tokyo: Shufunotomo, 1979.

Garnsey, Wanda, and Rewi Alley. *China , Ancient Kilns and Modern C eramics : A G uide to the P otteries*. Canberra and Miami, Fla.: Australian National University Press, 1983.

Gillette, Maris. "Copying, Counterfeiting, and Capitalism in Contemporary China: Jingdezhen's Porcelain Industry." *Modern China* 36, no. 4 (2010): 367–403.

Godden, Geoffrey A. *Oriental Export Market Porcelain and Its Influence on European Wares*. London: Granada, 1979.

Grave, Peter. "Melting Moments: Modelling Archeological High Temperature." In *New Directions in Archaeological Science*, edited by Andrew Fairbairn, Sue O'Connor, and Ben Marwick, 215-34. Canberra: ANU E Press, 2011.

Gunn, Geoffrey C. *History without Borders: The Making of an Asian World Region, 1000-1800*. Hong Kong: Hong Kong University Press, 2011.

Guy, John S. *Oriental Trade Ceramics in South-East Asia, Ninth to Sixteenth Centuries*. Oxford and NY: Oxford University Press, 1990.

Hackworth, Jason, and Niel Smith. "The Changing State of Gentrification." *Tijdschrift voor Economische en Sociale Geografie* 92, no. 4 (2001): 464–77.

Hall, Mark E. "Pottery Production during the Late Jomon Period: Insights from the Chemical Analyses of Kasori B Pottery." *Journal of Archaeological Science* 31 (2004): 1439–50.

Hang Seng and Jockey Club Creative Arts Centre. *Hang Seng Arts Empowerment Workshop Exhibition*. Hong Kong: Hang Seng and Jockey Club Creative Arts Centre, 2012.

Henderson, Joan. "Heritage, Identity and Tourism in Hong Kong." *International Journal of Heritage Studies* 7, no. 3 (2001): 219–35.

Hong Kong Baptist University (HKBU). "Jockey Club Creative Arts Centre." Accessed October 15, 2013. http://gao.hkbu.edu.hk/en/sps/jccac.html.

Hong Kong Jockey Club (HKJC). "Explore 'Old Factory, Emerging Arts' Exhibition at HK Racing Museum." December 16, 2008. Accessed December 13, 2012. http://corporate.hkjc.com/corporate/corporate-news/english/2008-12/news_2008121620748.aspx.

Hong Kong Museum of Art. "Chinese Ceramics and Seals Donated by the K.S. Lo Foundation." *Hong Kong Museum of Art Newsletter* 2 (April–June 2013): unpaginated.

———. *Cruising the Universe: Fantastic Animals in the Arts of China*. Hong Kong: Leisure and Cultural Services Department, 2013.

———. *Maritime Porcelain Road – Relics from Guangdong, Hong Kong and Macao Museums*. Hong Kong: Leisure and Cultural Services Department, 2013.

Hong Kong Wetland Park. "About Us – Background," 2012. Accessed September 5, 2012. http://www.wetlandpark.com/en/aboutus/index.asp.

Housing Authority. "Historical Background of Shek Kip Mei Estate." Accessed December 12, 2012. http://www.housingauthority.gov.hk/hdw/en/aboutus/events/community/heritage/about.html.

Hui, Desmond. "From Cultural to Creative Industries Strategies for Chaoyang District, Beijing." *International Journal of Cultural Studies* 9, no. 3 (2006): 317–31.

Hulu Culture. "JCCAC." Accessed December 13, 2012. http://www.huluhk.org/SSP/heritage.php?id=jccac&lang=en.

Hwang, Jim. "Going for Broke." Taiwan Info. Accessed November 3, 2013. http://taiwaninfo.nat.gov.tw/fp.asp?xItem=24038&CtNode=124.

Hyghe, Francois-Bernard. "The Kingdom of Silla and the Treasures of Nara (Ancient Korea Kingdom; Buddhist Temple in Nara, Japan)." *UNESCO Courier* (July 1991): 44–46.

Iaccarino, Ubaldo. "Manila as an International Entrepot: Chinese and Japanese Trade with the Spanish Philippines at the Close of the 16th Century." *Bulletin of Portuguese/Japanese Studies* 16 (2008): 71–81.

Idemitsu Museum of Arts. *Treasures along the Silk Road: Steppe and Oasis Routes/Sea Route*. Tokyo: Idemitsu Museum of Arts, 2001.

———. *Kakiemon to Nagashima* [Kakiemon and Nabeshima]. Tokyo: Idemitsu Bijutsukan, 2008.

Imai, Yuko. "Changes in French Tastes for Japanese Ceramics." *Japan Review* 16 (2004): 101–27.

Imakiire, Kana. "Japonisme in the 19th Century and Western Tableware Made in Japan." *Daisankai Kokusai Nihongaku Consortium*. Tokyo: Ochanomizu University, 2008.

Impey, Oliver. *The Early Porcelain Kilns of Japan: Arita in the First Half of the Seventeenth Century*. Oxford: Clarendon Press, 1996.

Information Services Department, Hong Kong Special Administrative Region Government. "Creative Industries." Accessed December 17, 2012. http://www.gov.hk/en/about/abouthk/factsheets/docs/creative_industries.pdf.

Itani, Yoshie. "Meijiki ni okeru Nihonjin no shokuseikatsu no hensen to yushutsujiki nitsuite" [Westernization of Japanese food culture and export porcelain in the Meiji era]. *Yūgō bunka kenkyū* [Cultural Fusion Studies] 6 (October 2005): 72–87.

Iten, Charly. "Ceramics Mended with Lacquer — Fundamental Aesthetic Principles, Techniques and Artistic Concepts." In *The Aesthetics of Mended Japanese Ceramics*, edited by Herbert F. Johnson Museum of Art and Museum fur Lackkunst. Ithaca: Herbert F Johnson Museum of Art, Cornell University, 2008.

Izushi, Hiro. "Conflict between Two Industrial Networks: Technological Adaptation and Interfirm Relationships in the Ceramics Industry in Seto, Japan." *Regional Studies* 31, no. 2 (18 August 2010): 117–29.

Jahn, Gisela. *Meiji Ceramics: The Art of Japanese Export Porcelain and Satsuma Ware*, 1868-1912. Translated from the German, edited by Michael Foster. Stuttgart: Arnoldsche, 2004.

John and Mable Ringling Museum of Art. *Crosscurrents of Design: Asian Export Ceramics at the Ringling A Resource Guide*. Florida: The John and Mable Ringling Museum of Art, n.d.

Jorg, Christiaan J. A. *Fine and Curious Japanese Export Porcelain in Dutch Collections*. Amsterdam: Hotei Publishing, 2003.

Katsuki, Hiroaki, Nobuaki Kamochi, Atsunori Shiraishi, Woo-Seok Cho, Jae-Hwan Pee, Kyung-Ja Kim, Eui-Seok Choi, and William M. Carty. "Some Properties of the Early Arita Celadon." *Journal of the Ceramic Society of Japan* 119, no. 1392 (August 2011): 672–76.

Kerr, Rose, and Nigel Wood. *Part 12: Ceramic Technology*. Science and Civilization in China 5. Cambridge: Cambridge University Press, 2004.

Kinloch, Valerie. "Youth Representations of Community, Art, and Struggle in Harlem." *New Directions for Adult and Continuing Education* 116 (2007): 37–49.

Kobayashi, Hitoshi. "History of Oriental Ceramics: The Evolution of Japanese Ceramics." The Museum of Oriental Ceramics, Osaka, 2009. Accessed March 15, 2012. http://www.moco.or.jp/en/intro/history_j.html.

Koyama, Fujio. *Nihon tōji no dentō* [Traditions of Japanese ceramics]. Kyōtō: Tankō shinsha, 1967.

Kyodo. "Arita Kicks Off Annual Ceramics Fair." *Japan Times*, April 30, 2001. Accessed July 31, 2013. http://www.japantimes.co.jp/news/2001/04/30/national/arita-kicks-off-annual-ceramics-fair/#.UfkWvV_yyUk.

Kyushu Ceramic Museum, *Koimari Nyūmon* [Introduction to ceramic ware "Koimari"]. Kyoto: Seigensha, 2010.

Law, Roger et al. *Porcelain City: Jingdezhen.* London: Victoria and Albert Museum, 2012.

Lazzaretti, Luciana. "City of Art as a High Culture Local System and Cultural Districtualization Processes: The Cluster of Art Restoration in Florence." *International Journal of Urban and Regional Research* 27, no. 3 (2003): 635–48.

Le French May Organisation. *Le French May 17 April–23 June 2013 French Arts Festival Hong Kong & Macau*. Hong Kong: Quality Printing Limited, 2013.

Lee, Hun-Chang. "When and How did Japan Catch Up with Korea?: A Comparative Study of the Pre-industrial Economies of Korea and Japan." Center for Economic Institutions Working Paper Series, No. 2006–15, Institute of Economic Research, Hitotsubashi University, February 2002.

Ley, David. "Artists, Aestheticisation and the Field of Gentrification." *Urban Studies* 40, no. 12 (2003): 2527–44.

Lin, Renchuan. "Fukien's Private Sea Trade in the 16th and 17th Centuries." In *Development and Decline of Fukien Province in the 17th and 18th Centuries*, edited by Eduard B. Vermeer, 163–215. Leiden; New York: Brill, 1990.

Lin, Wei-sheng. "The VOC Trade in Taiwan." *IIAS Newsletter: Research and Report* 31 (July 2003): 30.

Little, Stephen. "Economic Change in Seventeenth-Century China and Innovations at the Jingdezhen Kilns." *Ars Orientalis* 26 (1996): 47–54.

Major, John. "Geographical Setting of the Silk Roads." Asia Society, August 18, 2008, http://asiasociety.org/countries-history/trade-exchange/geographical-setting-silk-roads?page=0%2C3.

Mak, Yee Fun. "About Me." Accessed October 28, 2012. http://makyeefun.com/aboutme.html.

Maske, Andrew. "A Brief History of Takatori Ware." The University of North Carolina at Chapel Hill. Accessed March 15, 2012. http://www.unc.edu/~mpitelka/takatori/index.html.

McCarthy, John. "Regeneration of Cultural Quarters: Public Art for Place Image or Place Identity?" *Journal of Urban Design* 11, no. 2 (2006): 243–62.

McKercher, Bob, Pamela S.Y. Ho, and Hilary du Cros. "Relationship between Tourism and Cultural Heritage Management: Evidence from Hong Kong." *Tourism Management* 26, no. 4 (2004): 539–48.

Medley, Margaret. *The Chinese Potter: A Practical History of Chinese Ceramics*. London and New York: Phaidon Press, 2006.

Moeran, Brian. "Japanese Social Organization and the Mingei Movement." *Pacific Affairs* 54, no. 1 (1981): 42–56.

———. *Lost Innocence: Folk Craft Potters of Onta, Japan*. Berkeley: University of California Press, 1984.

———. *Folk Art Potters of Japan: Beyond Anthropology of Aesthetics*. Great Britain: Curzon, 1997.

Moll-Murata, Christine. "Guilds and Apprenticeship in China and Europe: The Ceramics Industries of Jingdezhen and Delft." Paper presented to the S.R. Epstein Memorial Conference: "Technology and Human Capital Formation in the East and West." June 18–21, 2008. http://www2.lse.ac.uk/economicHistory/Conferences/Epstein%20Memorial%20Conference/PAPER-MollMurata.pdf.

Montlake, Simon. "Clinton Visit to Myanmar Stokes Interest in Asia's Last Frontier." *Forbes/Wall Street Journal*, November 30, 2011. Accessed September 19, 2012. http://www.forbes.com/sites/simonmontlake/2011/11/30/clinton-visit-to-myanmar-stokes-interest-in-asias-last-frontier/.

Morimoto, Asako. "A New Approach to Enigmas from Medieval Hakata: Trade Ceramics as Seen from Archaeological Data." *Guoli Taiwan daxue meishushi yanjiu jikan* [Taida Journal of Art History] 7 (1999): 89–90.

Nagatake, Takeshi. *Arita yakimono no dokuhon* [Publication on Arita Pottery]. Japan: Aritatōji Bijutsukan, 1961.

———. *Nihon no tōjiki: Sono kamaba o tazunete* [Japanese ceramics: Visit to the Kilns]. Tokyo: Shakai Shisōsha, 1966.

———. *Classic Japanese Porcelain: Imari and Kakiemon*. Tokyo, New York and London: Kodansha International, 2003.

Nagumo, Ryū. *Tōgei* [Ceramics]. Tokyo: Nichibō Shuppansha, 1998.

Nan, Kyi Kyi Khai. "Myanmā bōeki tōjiki no kenkyū: Marutaban tsubo to hakuyūzara o chūshin ni" [The study on Myanmar trade ware: Martaban Jar and white dish]. Kanazawa daigaku kōkogaku kiyō 30 (March 2009): 101–36.

New Taipei City Yingge Ceramics Museum. "Yingge Wares – Yingge Ceramics Industry Rejuvenate." Accessed November 2, 2013. http://exhibition.ceramics.ntpc.gov.tw/yingge-wares/en/index_2.html.

Nguyen, Thi Ha Thanh. "European Trade on the Far East and the Mercantile Relationship with Vietnam from the 16th to 19th Century." *Higashi Ajia bunka kōshō kenkyū* [Journal of East Asian Cultural Interaction Studies] 2 (March 2009): 353–66. http://www.icis.kansai-u.ac.jp/data/journal02-v1/24_nguen.pdf.

Noda, Kōichi. *Hajimeteno tōgeigama erabi* [First time to choose kilns]. Tokyo: Seibundō, 2005.

Nogami, Takenori. "On Hizen Porcelain and the Manila-Acapulco Galleon Trade." *Bulletin of the Indo-Pacific Prehistory Association* 26 (2006): 124–130.

Ōba, Kōji. *Chūsei Nihon saidai no bōeki toshi* [The biggest trading city in Japan during the Medieval Period]. Tokyo: Shinsensha, 2009.

Ōhashi, Kōji. *Umi o watatta koimari—seramikku rōdo* [Exported old-Imari ceramic road]. Kyoto: Seigensha, 2011.

Oriental Ceramic Society of Hong Kong Limited and Art Museum, The Chinese University of Hong Kong. *Divine Power—The Dragon in Chinese Art*. Hong Kong: The Oriental Ceramic Society of Hong Kong Limited and Art Museum, The Chinese University of Hong Kong, 2012.

Outdoor Life Books Co. Ltd. Tour Information. *Yingge, Sanxia, Ulai, Beiyi Gonglu*. Taiwan: Outdoor Life, 2009.

Pelehach, Patricia. *Dancing with Fire: Building an Anagama Kiln and an Artistic Community*. Pennsylvania: Catskill Arts Society, 2006. Accessed March 15, 2012. http://moca-ny.org/learnKiln.html.

Pieterse, Jan Nederveen. "Global Multiculture: Cultures, Transnational Culture, Deep Culture." In *Hybrids, Differences, Visions: On the Study of Culture*, edited by Claudio Baraldi, Andrea Borsari, and Augusto Carli, 1-14. Aurora, CO: Davies Group, 2010. http://jannederveenpieterse.com/pdf/Global_multi.pdf.

Pollard, Clare. *Master Potter of Meiji Japan: Makuza Kōzan (1842–1916) and his Workshop*. Oxford: Oxford Oriental Monographs, 2001.

Pottery Workshop. "Zheng Yi Trading Company." Accessed October 30, 2013. http://www.potteryworkshop.com.cn/trading.asp?183.html.

Rawski, Evelyn S., and Thomas G. Rawski. "China's Economic Development and Global Interactions in the Long Run." Paper prepared for the Harvard-Hitotsubashi-Warwick Conference "Economic Change around the Indian Ocean in the Very Long Run," Venice, July 22–24, 2008. http://www2.warwick.ac.uk/fac/soc/economics/news/forums/conferences/econchange/programme/rawski_-_venice.pdf.

Ren, Seikichi. "Higashi Ajia no bunka kankyō no keisei" [Cultural environmental formation in East Asia]. In *Kankyō to bunka – (Bunka kankyō) no shosō* [Environment and culture—aspects of cultural environment], edited by Nagasaki Daigaku Bunka Kankyō Kenkyūkai, 283-306. Kyushu: Kyushu University Press, 2000.

Rotondo-McCord, Lisa, et al. *Imari: Japanese Porcelain for European Palaces: From the Freda and Ralph Lupin Collection*. New Orleans New Orleans Museum of Art, 1999.

Rousmaniere, Nicole Coolidge. *Vessels of Influence: Chinese Ceramics Imported into Japan and the Formation of the Porcelain Industry*. Bristol: Bristol Classical Press, 2012.

Saga Kenritsu Kyūshū Tōji Bunkakan. *Earth and Fire: The Historical Development of Kyushu Ceramics: The Kyushu Ceramic Museum, The Permanent Exhibition Guide Book*. Saga, Arita: Saga Kenritsu Kyūshū Tōji Bunkakan, 1996.

Sakai, Takashi. "Indoyō no tōji bōeki: Toruko to Higashi Ajia no kōryū o megutte" [The ceramic trade of the Indian Ocean: Concerning exchange between Turkey with the Eastern Part of Asia]. *Jōchi Ajia gaku* [The Journal of Sophia Asian Studies] 23 (2005): 261–309.

Sakuraba, Miki. "Evolution of Porcelain Tea Ware in Europe From Hizen Porcelain Tea Ware to European Porcelain Tea Service." Translated by Jenine Heaton. In *Cultural Reproduction on its Interface: From the Perspectives of Text, Diplomacy, Otherness, and Tea in East Asia*, edited by Nishimura Masanari et al. The International Academic Forum for the Next Generation Series 1. Osaka: Institute for Cultural Interaction Studies, Kansai University, 2010.

Sanger, David. "Arita Journal; At Hallowed Pottery, the Master's Wheel Is Stilled." *New York Times*, May 1, 1990. Accessed March 15, 2012. http://www.nytimes.com/1990/05/01/world/arita-journal-at-hallowed-pottery-the-master-s-wheel-is-stilled.html?pagewanted=all&src=pm.

Santagata, Walter. "Cultural Districts, Property Rights and Sustainable Economic Growth." *International Journal of Urban and Regional Research* 26, no. 1 (2002): 9–23.

Sasaki, Tatsuo. "Ceramic Trade via the Oceans." *The Archaeological Journal of Kanazawa University* 67 (2010): 27–32.

Schiffer, Nancy N. *Japanese Export Ceramics 1860-1920*. Pennsylvania: Schiffer Publishing Ltd., 2000.

Seattle Art Museum. *Guide to the Porcelain Room: Seattle Art Museum*. Oregon: CDS Publications, 2007.

Seto Municipal Center of Multimedia and Traditional Ceramics. "About Seto Sometsuke (Seto)." Accessed March 12, 2012. http://www.city.seto.aichi.jp/e/tourist_infomation/municipal/06.html.

Seyock, Barbara. "Trade Ceramics from the Goto Islands (Japan), Circa Sixteenth to Early Seventeenth Century: The Yamami Underwater Site (Ojika) and Related Issues." *Asian Perspectives* 46, no. 2 (Fall 2007): 335–60.

Shi, Yongan. *Riben gutaoci*. Changchun: Jilin Meishu Chubanshe, 1992.

Shimura, Goro. *The Story of Imari: The Symbols and Mysteries of Antique Japanese Porcelain*. Berkeley California: Ten Speed Press, 2008.

Soejima, Kiyoshi, and Kiyoshi Tomimatsu. "Research on Firing Distortion Prediction and Correction Techniques for Ceramics Design." In *Processing and Properties of Advanced Ceramics and Composites II*, vol. 220, edited by Narottam P. Bansal, Jitendra P. Singh, Jacques Lamon, Sung R. Choi, and Morsi M. Mahmoud, 159–72. Hoboken, NJ: John Wiley & Sons, 2010. doi: 10.1002/9780470930953.ch17.

SOIL. "About SOIL." Accessed August 2, 2013. http://craftbysoil.wordpress.com/about/.

———. "Burma: The Curse of Independence by Shelby Tucker, 2001." Craft by Soil, September 9, 2012. Accessed September 20, 2012. http://craftbysoil.wordpress.com/2012/09/09/burma-the-curse-of-independence-by-shelby-tucker-2001/.

SOIL's Facebook page. September 16 2012. Accessed September 20, 2012. http://www.facebook.com/craftbysoil.

Stevenson, John, and John Guy. *Vietnamese Ceramics: A Separate Tradition*. Chicago, IL: Art Media Resources with Avery Press, 1997.

Stites, Richard. "Small-Scale Industry in Yingge, Taiwan." *Modern China* 8, no. 2 (1982): 247–79.

Stitt, Irene. *Japanese Ceramics of the Last 100 Years*. New York: Crown Publishers, 1974.

Thong, Quach Huy, and Ranjith Perera. "Reduction of Environmental Pollution in Handicraft Manufacturing Enterprises by Using Cleaner Technology: A Case Study of the Traditional Ceramic Producing Village in Bat Trang, Hanoi, Vietnam." In *Proceedings: Sharing of Experiences on Air Pollution Abatement in Southeast Asia (Pre-event, Better Air Quality 2006 Workshop)*, Indonesia, Yogyakarta, December 12, 2006, 61-71. Pathumthani: Asian Institute of Technology, 2008.

Time Out Hong Kong. "Four Things We Love About...Sau Wa Fong." Discover Hong Kong, Hong Kong Tourism Board, May 23, 2012. Accessed September 20, 2012. http://www.discoverhongkong.com/nz/shop/whats-hot/sau-wa-fong.jsp.

Tōkyō Kokuritsu Hakubutsukan [Tokyo National Museum]. *Kareinaru Imari, miyabi no Kyōyaki* [Splendid and refined Imari ware and Kyoto ware ceramics]. Tokyo: Yomiuri Shinbunsha, 2005.

Tolliday, Steven, and Yasushi Yonemitsu. "Microfirms and Industrial Districts in Japan: The Dynamics of the Arita Ceramic-ware Industry in the Twentieth Century." *Journal of Japanese Studies* 33, no. 1 (2007): 29–66.

Uchida, Hoshimi. *Short History of the Japanese Technology*. Mitaka, Tokyo: The History of Technology Library, 1995. http://www.ied.co.jp/isan/sangyo-isan/uchida.PDF.

Ueki, Mikio. "The Modernization and Japanization of Contemporary Japanese Ceramics." *Journal of Hokkai-Gakuen University* 123 (March 2005): 1-8. Accessed March 12, 2012. http://hokuga.hgu.jp/dspace/bitstream/123456789/1108/1/GAKUEN-123-1.pdf.

VOC Arita Listed Company. *The Hague city Art Museum and Nishi Nippon Shimbunsha, Arita Delft China*. Japan: VOC Arita Listed Company, 1993.

Volk, Alicia. *In Pursuit of Universalism*. California: UC Press, 2010.

Wang, Xiaoying. "Hong Kong, China, and the Question of Postcoloniality." In *Postmodernism and China*, edited by Arif Dirlik and Xudong Zhang, 89–119. Durham: Duke University Press, 2000.

Warmelo, W. Van. "The Development of Small Enterprises in Taiwan, Republic of China." *Industry of Free China* 28, no. 5 (1967): 2–21.

Watt, George. "Burmese Lacquerware and Burmese Varnish. (*Melanorrhoea usitata*, Wall)." *Bulletin of Miscellaneous Information (Royal Gardens, Kew)* 1906, no. 5 (1906): 137–47.

Wood, Nigel. "New Discoveries in Chinese Ceramics." Accessed October 29, 2013. http://www.haughton.com/system/files/articles/2010/01/27/102/icf_s_1997_4th.pdf.

Yamaguchi Prefecture Shunan Industrial Company Database. "Yoshida-ya shōyu," 2011. Accessed February 15, 2013. http://www.shunan-marketing.jp/detail.php?cci=1&manage=2412.

Yamashita, Eiichi. "Ceramics: The Seto Ceramics School." In *Vocational Education in the Industrialization of Japan*, edited by Toshio Toyoda, 93-108. Tokyo: United Nations University Press, 1987.

Yao, Chun. "Tanjiu Riben taoci yaolu jishu" [Investigating Japan's Kiln Techniques]. *Zhongguo taoci* [Chinese Ceramics], February 23, 2011. http://enewspaper.ju51.com/xbcy/XBTC/data/20110223/html/30/contentpage_1_0.html.

Yim, Wai Wai. "Yim Wai-wai – [Art] Attack 16." Last modified August 18, 2009. Accessed September 5, 2012. http://www.choochootv.com/choochootv-shows-/artattack/195-yim-wai-wai-artattack-16.

———. *LvLn Virus*. Hong Kong: iSquare Design and Development Ltd, 2011.

———. *Front Side Gallery Invitation.* Yim Wai Wai's Solo Exhibition pamphlet. Hong Kong, December 9, 2012.

———. "'LvLn Virus' Ceramics Installation by Yim Wai Wai." Accessed December 13, 2012. http://www.youtube.com/watch?v=1oZjkrUMmIA.

"Yingge shao" [Yingge Ware]. *Golden Age Magazine*. Hong Kong: Golden Age/Senior Citizen Home Safety Association, 2012.

Yoshioka, Yasunobu. *Tōjiki no sekai* [The world of porcelain]. Tokyo: Yamakawa Shuppansha, 2001.

Yuet Tung China Works. "About Us." Accessed September 5, 2012. http://www.porcelainware.com.hk/aboutus.html.

Yung, Vanessa. "Myanmar Pops Up Here." *South China Morning Post* (SCMP), April 29, 2012. Accessed September 20, 2012. http://www.scmp.com/article/999498/myanmar-pops-here.

www.ingramcontent.com/pod-product-compliance
Lightning Source LLC
LaVergne TN
LVHW101447200926
848483LV00001B/71